COMPUTATIONAL BIOLOGY OF CANCER

LECTURE NOTES AND MATHEMATICAL MODELING

COMPUTATIONAL BIOLOGY OF CANCER

LECTURE NOTES AND MATHEMATICAL MODELING

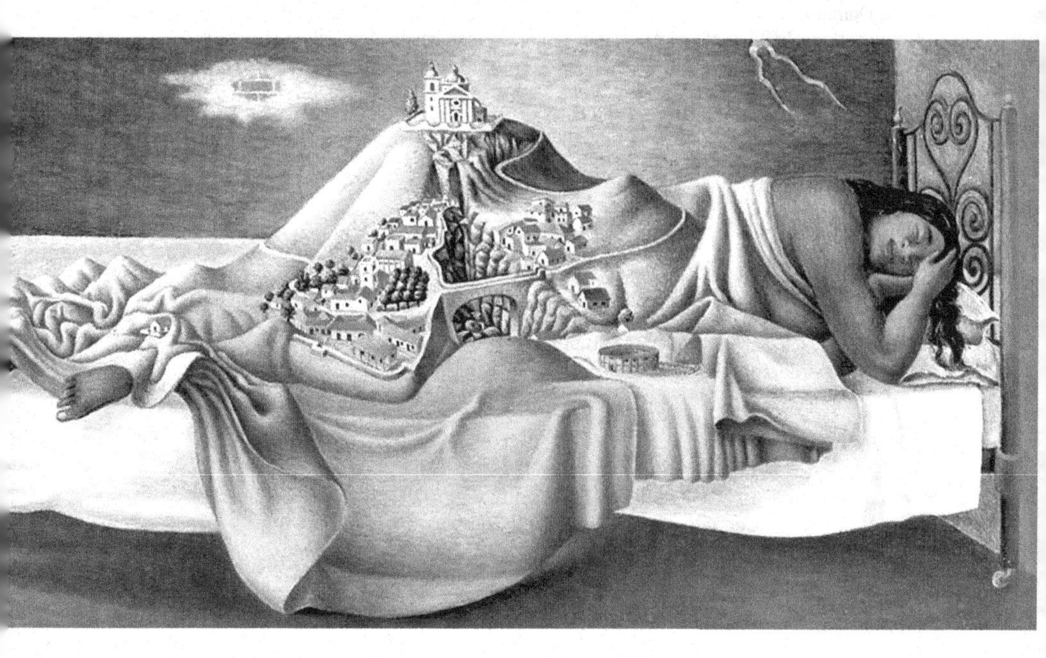

Dominik Wodarz
University of California, Irvine, USA

Natalia L. Komarova
University of California, Irvine, USA
and
Rutgers University, USA

NEW JERSEY • LONDON • SINGAPORE • BEIJING • SHANGHAI • HONG KONG • TAIPEI • CHENNAI

Published by

World Scientific Publishing Co. Pte. Ltd.

5 Toh Tuck Link, Singapore 596224

USA office: 27 Warren Street, Suite 401-402, Hackensack, NJ 07601

UK office: 57 Shelton Street, Covent Garden, London WC2H 9HE

Library of Congress Cataloging-in-Publication Data

Wodarz, Dominik.

Computational biology of cancer : lecture notes and mathematical modeling / Dominik Wodarz and Natalia Komarova.

p. cm.

Includes bibliographical references and index.

ISBN-13 978-981-256-027-8 -- ISBN-10 981-256-027-0 (alk. paper)

1. Cander--Mathematical models. I. Komarova, Natalia. II. Title.

RC267 .W535 2005

616.99'4'00724--dc22 2004056774

British Library Cataloguing-in-Publication Data

A catalogue record for this book is available from the British Library.

Image on the cover: Antonio Ruiz, El Sueño de la Malinche
1939, Oil on canvas, 29.5 × 40 cms.
Courtesy of Galería de Arte Mexicano

First published 2005
Reprinted 2008

Printed in Singapore

Dedication Page

To our families:
Klara, Hans-Walter, Andi and Pinktche
Nadezhda, Leonid and Yozh.

Preface

People who are not mathematicians are strange. At least, they think differently. Things that we take for granted are exciting news for them. Problems which are most interesting for us don't even get registered with them. And, most annoyingly, they have a habit of asking really difficult questions. For instance, you are working on a model of colon cancer initiation. The biologist keeps asking "Can you include this? Can you include that in your model?" You smile meekly ("It's *hard*!"), and then, just to finish you off, he adds: "By the way, things work differently whether it is in the front or on the back of the colon." You didn't tell him that you modeled the colon as a *sphere*...

Something has to change. Many mathematical papers have been written about cancer, many interesting models created, many challenging questions asked. However, theoretical work is only valuable to the field of cancer research if models are validated by experiment, predictions are tested, and models are revised in the light of empirical data. Such an integrated and multi-disciplinary approach is so far lacking in the context of carcinogenesis. Theorists sometimes do modeling for the sake of the mathematical analysis that they can successfully pursue, which is of zero relevance to the field of cancer research. This creates general skepticism among the experimentalists. On the other hand, experimental biologists are often unfairly dismissive of the role of theory. It is not uncommon to hear that all theory is naive and that theoretical biologists cannot possibly grasp the full complexity of the biological reality, leave alone modeling it accurately.

The broad aim of this book is to provide an introduction to mathematical modeling in cancer research, and we hope that this will contribute to bridging the gap between mathematical modelers and experimental oncologists. The book is written with this goal in mind. On the one hand we will introduce the mathematical methodology which underlies the theoretical work. On the other hand, we discuss how the modeling results can help us generate new biologically relevant insights, interpret data, and design further experiments.

Which readership do we have in mind? On a sunny day we think that the book will be read with enthusiasm by both applied mathematicians who wish to learn about theoretical work in cancer, and by experimental oncologists who would like to bring new, interdisciplinary dimensions to their research. On a rainy day, however, we realize that this is easier said than done. The book is certainly suited very well for applied mathematicians, because they are already familiar with the backbone of theoretical biology: computation and mathematics. We hope that after reading this book, they will sense a longing to learn much more about cancer biology and to pursue modeling work which is closely linked to biological data. The book should also be relatively easy to digest by biologists who will understand all the concepts even if they might not be familiar with some of the math. It will motivate them to get more familiar with mathematical methods and to widen their horizons; or, in case of experimentalists, it will allow them to see that modeling can give rise to interesting concepts which could help them formulate new questions and experiments. The biggest challenge, of course, are biologists who close the book when they encounter an equation, or who have the preconception that all theoretical work is naive and useless. Having many more sunny than rainy days in California, however, we have an optimistic outlook. We think that if enough experimental oncologists become enthusiastic about collaborating with modelers, theoretical work will spread through the community and bring many interesting results.

Besides the weather, our optimism is also fueled by experiences from another biomedical discipline: the interaction between pathogens and the immune system. Before 1990, all of the research which was considered "biologically relevant" by immunologists and virologists was experimental, and theoretical work was met with great skepticism. Subsequently, a wave of interesting work emerged which involved collaborations between some experimental immunologists/virologists and mathematical modelers. In fact, one of the most influential and widely cited papers in AIDS research, which appeared in a couple of Nature papers in 1995, came about through collaborations between modelers and experimental labs. Today, theory plays a relatively large role in immunology, to the extent that experimental design and research directions can be influenced by results obtained from mathematical modeling.

Why has mathematical modeling become an integral part of immunological research? The interactions between pathogens and the immune system involves many different components which interact with each other. These interactions are highly complex, non-linear, and can result in counter-intuitive outcomes. People started to realize that simple verbal or graphical reasoning is not sufficient to obtain a complete understanding of these interactions. Instead, it became clear that mathematical models are essential. They provide a solid framework upon which to generate hypotheses, interpret data, and design new experiments.

Cancer research is similar in this respect. It involves multiple interactions between molecules, cells, and their environment. As in immunology, we expect that mathematical models are essential to complement experimental work in order to obtain a satisfactory understanding of this complex biological system. We hope that our book will help to push the field of cancer research a little bit in the direction in which immunology has developed over time.

This brings up an important question. Does this book cover all aspects of cancer research? Certainly not! In fact, this would be impossible, unless you write many many volumes. Cancer is a very complicated topic and can be studied on many different levels. We chose to focus on one particular aspect of cancer: the process of carcinogenesis as *somatic evolution* of cells. This is a suitable topic for the introduction of mathematical modeling. Besides being deeply rooted in cancer biology, it is also partly based on the principles of evolutionary theory (mutation and selection) - a field where mathematical modeling has played a significant role since the early 1930s.

We would like to thank a number of people who got us fascinated by cancer research and who ultimately enabled us to write this book. These are experimental oncologists who are already very open minded towards computational approaches, who are willing to discuss theoretical ideas, and who educate us about the field. Arnie Levine provided the initial stimulus which got us working on cancer. At the Institute for Advanced Study in Princeton, where this book was partly written, we enjoyed many interesting and important discussions. Rick Boland at Baylor University, Dan Gottschling, Lee Hartwell and Chris Kemp at the Hutch, Larry Loeb at the University of Washington, and Vladimir Mironov at Medical University of South Carolina have provided many useful discussions and insights which were essential for our modeling work.

We are also grateful to our fellow theoreticians. In our old Princeton group, Steve Frank (also at UCI), Yoh Iwasa, David Krakauer, Alun Lloyd,

Martin Nowak, Karen Page, and Joshua Plotkin. In the Hutch, Mark Clements, Bill Hazelton, Georg Luebeck, and Suresh Moolgavkar. At Rutgers, Liming Wang, Eduardo Sontag, and the math physics group. Special thanks to Victoria Kamsler for artistic advice, Michel Reymond for providing food for inspiration, and Diane Depiano, Susan Higgins, and Anne Humes for always being there for us.

We are indebted to the institutions at which we worked while writing the book: Fred Hutchinson Cancer Research Center, University of California Irvine, Rutgers University, and Institute for Advanced Study in Princeton. Finally we would like to thank Steve Frank, Francisco Ayala, and the Biology and Math Departments of the University of California Irvine for recruiting us recently. They gave us an excellent work environment, and allowed us to be at a place with many sunny days.

Dominik Wodarz and Natalia L. Komarova,
Princeton & Irvine, 2004

Contents

Chapter 1

Cancer and somatic evolution

1.1 What is cancer?

The development and healthy life of a human being requires the cooperation of more than ten million cells for the good of the organism. This cooperation is maintained by signals and cellular checkpoints which determine whether cells divide, die, or differentiate. The phenomenon of cancer can be defined on various levels. On the most basic level, cancer represents the collapse of this cooperation. This results in the selfish, uncontrolled growth of cells within the body which eventually leads to the death of the organism. The first chapter will discuss several aspects of cancer biology. This forms the background for the mathematical models which are presented in this book. Of course, cancer biology is a very complicated topic and involves many components which are not mentioned here. A comprehensive review of cancer biology is given in standard textbooks, such as [Kinzler and Vogelstein (1998)].

It is commonly thought that cancer is a disease of the DNA. That is, uncontrolled growth of cells is the result of alterations or mutations in the genetic material. More precisely, the emergence of cancer may require the accumulation of multiple mutations which allow cells to break out of the regulatory networks which ensure cooperation. This concept is referred to as *multi-stage carcinogenesis*. Once a cancerous cell has been created it can undergo a process known as *clonal expansion*. That is, it gives rise to descendants by cell division, and the population of cells grows to higher numbers. During this process, cells can acquire a variety of further mutations which leads to more advanced progression. A cancer is typically comprised of a variety of different genotypes and represents a "mosaic" of cell lineages. The growth of a single, or *primary*, cancer does not usually lead to the death of the organism. Some cancer cells can, however, acquire the ability to enter the blood supply, travel to a different site, and start growing in a different organ. This process is referred to as *metastasis.* It is usually the metastatic growth which kills the organism.

1.2 Basic cancer genetics

Specific genes ensure that the integrity of cells is maintained and that uncontrolled growth is prevented. When these genes are mutated, cells become prone to developing a cancerous phenotype (also referred to as *transformation*). These genes can be broadly divided into three basic categories [Vogelstein *et al.* (2000a)]: oncogenes, tumor suppressor genes, and repair genes.

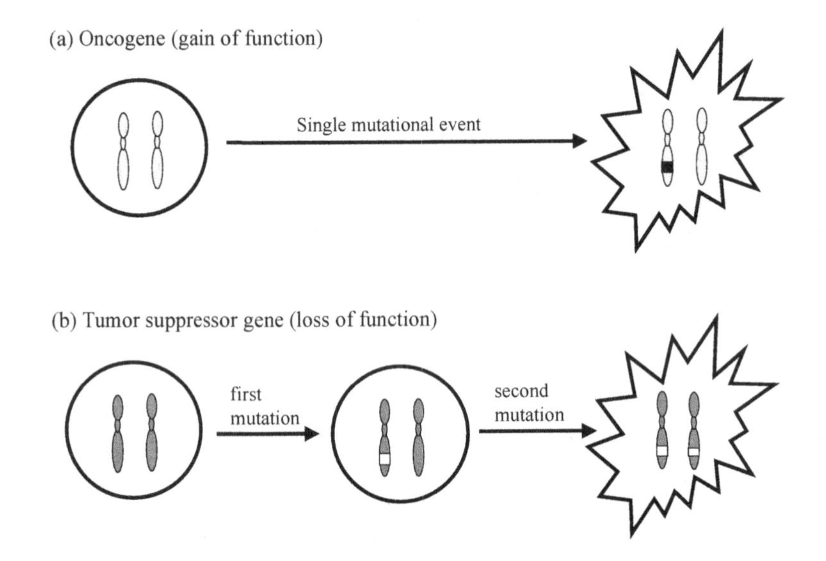

Fig. 1.1 The concept of (a) oncogenes and (b) tumor suppressor genes. Oncogenes result in a gain of function if one of the two copies receives an activating mutation. Tumor suppressor genes can be inactivated (loss of function) if both copies are mutated.

In healthy cells, oncogenes (Figure 1.1) promote the regulated proliferation of cells in the presence of the appropriate growth signals. The best example is the renewal of epithelial tissue such as the skin or the lining of the gastrointestinal tract. When oncogenes become mutated they induce the cell to divide continuously, irrespective of the presence or absence of growth signals. This can result in unwanted growth and cancer. Examples of oncogenes include Ras in colon cancer or BCL-2 in lymphoid cancers. Only a single mutation is required to activate an oncogene because it causes a "*gain of function*". Normal cells have two copies of every gene and chromosome; one derived from the mother, the other derived from the father. If any of the copies becomes activated, the cell attains the new behavior.

Tumor suppressor genes (Figure 1.1), on the other hand, are responsible for stopping growth in normal cells. Cell growth has to be stopped if a cell becomes damaged or mutated, or if cell death is required for normal tissue homeostasis. This is done either by preventing the cell from completing the cell cycle (*cell cycle arrest* or *senescence*), or by inducing a cellular program which results in cell death (*apoptosis*). In this way, altered cells cannot succeed to grow to higher levels and cannot induce pathology. When tumor suppressor genes become inactivated, the growth of altered cells is not prevented anymore, and this promotes the development of cancer. Because this type of gene needs to be inactivated rather than activated (i.e. a *loss of function event*), both the paternal and the maternal copies of the gene have to be mutated. Therefore, two mutational events are required for the inactivation of tumor suppressor genes. Because many cancers are initiated via the inactivation of a tumor suppressor gene, it is thought that cancer initiation often requires two hits. This idea was first formulated by Alfred Knudson and is called the "two hit hypothesis". Examples of tumor suppressor genes are the gene which encodes the retinoblastoma protein and which is inactivated in retinoblastomas, APC which is inactivated in colon cancer, and p53 which is inactivated in more than 50% of all human cancers.

Finally, repair genes are responsible for maintaining the integrity of genomes. When DNA becomes damaged, for example through the exposure to UV radiation or carcinogens contained in food, those genes make sure that the damage is removed and the cell remains healthy. If repair genes become mutated, cells can acquire new genetic alterations at a faster rate, and this promotes the process of carcinogenesis. For example, mutations in oncogenes or tumor suppressor genes are generated faster. Cells which have mutated repair genes are sometimes referred to as "mutator phenotypes" or "genetically unstable cells". Examples of repair genes are mismatch repair genes and nucleotide excision repair genes. Their inactivation promotes a variety of cancers. Loss of repair function usually requires two hits, although a single mutation might result in reduced function in the context of certain repair genes.

1.3 Multi-stage carcinogenesis and colon cancer

Cancer initiation and progression requires the sequential accumulation of mutations, most importantly in tumor suppressor genes and in oncogenes. The case study where this is understood in most detail so far is colorectal cancer. The colon consists of a collection of so-called *crypts*.

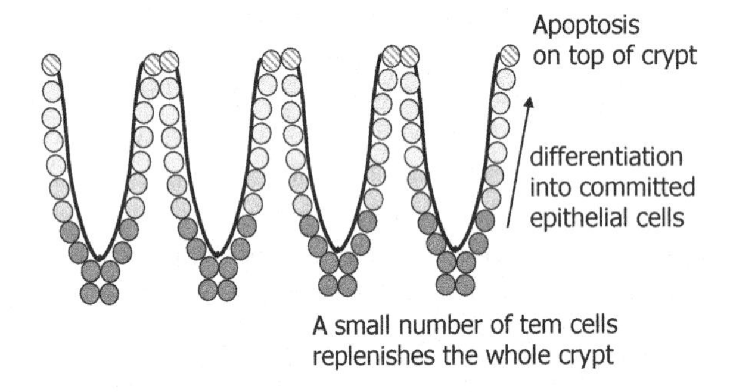

Fig. 1.2 Schematic diagram of crypts in the colon.

Crypts are involutions of the colonic epithelium (Figure 1.2). Stem cells are thought to be located at the base of the crypts. These are undifferentiated cells which can keep dividing and which give rise to differentiated epithelial cells. It is thought that stem cells divide asymmetrically. That is, stem cell division creates one new stem cell and one cell which embarks on a journey of differentiation. The differentiating cells travel up the crypt, perform their function, and die by apoptosis after about a week. Because the epithelial cells are relatively short lived, stem cell division has to give rise to new differentiated cells continuously in order to replenish the tissue. For this process to function in a healthy way, it is crucial that the differentiated cells die by apoptosis. If this cell death fails, we observe an accumulation of transformed cells around the crypts, and this gives rise to a mass of cells called a *dysplastic crypt* (Figure 1.3).

This is the first stage of colon cancer. In molecular terms, the death of differentiated cells is induced by the APC gene. APC is a tumor suppressor gene. Data suggest that the majority of colon cancers are initiated through

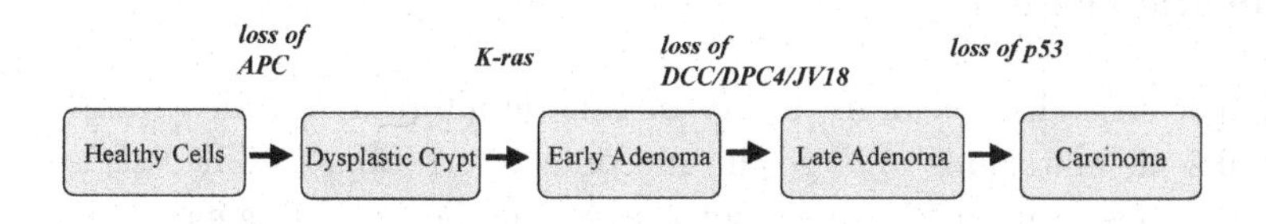

Fig. 1.3 Diagram describing the multi-stage progression of colon cancer. Drawn according to [Kinzler and Vogelstein (1998)].

the inactivation of the APC gene (Figure 1.3). A dysplastic crypt is also sometimes referred to as a polyp. As a subsequent step, many colon cancers activate the oncogene K-ras which allows the overgrowth of surrounding cells and an increase in the size of the tumor. This stage is called the early adenoma stage (Figure 1.3). In more than 70% of the cases, this is followed by the loss of chromosome 18q which contains several tumor suppressor genes including DDC, DPC4, and JV18-1/MADR2. This results in the generation of late adenomas (Figure 1.3). In the further transition from late adenoma to the carcinoma stage, p53 is typically lost in more than 80% of the cases (Figure 1.3). Further mutations are assumed to occur which subsequently allow the colon cancer cells to enter the blood system and metastasize. Note that this sequence of event is not a hard fact, but rather a caricature. The exact details may vary from case to case, and new details emerge as more genetic research is performed.

This is a clear example of cells acquiring sequential mutations in a multi-step process while they proceed down the path of malignancy. This gives rise to an important question. The multi-step process requires many mutations. The inactivation of each tumor suppressor gene requires two mutations, and the activation of each oncogene requires one mutation. The physiological mutation rate has been estimated to be 10^{-7}per gene per cell division. Is this rate high enough to allow cells to proceed through multi-stage carcinogenesis during the life time of a human? Some investigators argue that the process of clonal expansion involves a sufficient number of cell divisions in order to account for the accumulation of all the mutations. A competing argument says that the accumulation of the oncogenic mutations requires a loss of repair function and the generation of mutator phenotypes (i.e. genetically unstable cells). Genetic instability is a defining characteristic of many cancers. It is reviewed in the following section.

1.4 Genetic instability

Many cancer cells show a large variety of genetic alterations which range from small scale mutations to large chromosomal aberrations. While this is an intriguing observation, this does not prove that the cells are genetically unstable. The alterations could come about through a variety of factors, such as the exposure to extensive damage at some point in time, or specific selective conditions. Genetic instability is defined by an increased *rate* at which cells acquire genetic abnormalities [Lengauer *et al.* (1998)]. That is, cells have a defect in specific repair genes which results in higher variability. Indeed, studies have shown that many cancer cells are characterized by an increased rate at which genetic alterations are accumulated and are truly genetically unstable. Different types of genetic instabilities can be distinguished. They can be broadly divided into two categories. Small sequence instabilities and gross chromosomal instabilities (Figure 1.4).

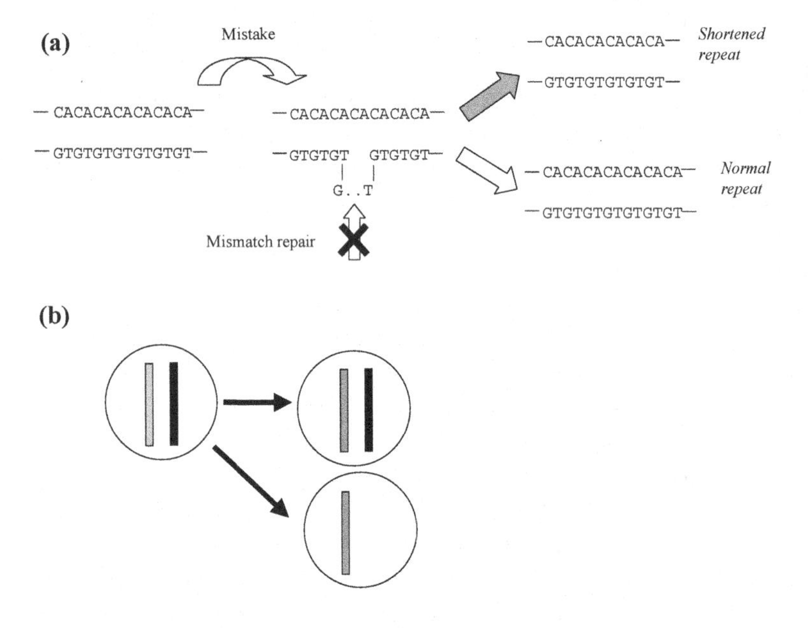

Fig. 1.4 Schematic diagram explaining the concept of genetic instability. (a) Small scale instabilities, such as MSI, involve subtle sequence changes. With MSI, mismatch repair genes are defect and this leads to copying mistakes in repeat sequences. (b) Chromosomal instability involves gross chromosomal changes, such as loss of chromosomes.

Small sequence instabilities involve subtle genetic changes which can dramatically speed up the process of cancer progression. Defects in mismatch repair mechanisms give rise to microsatellite instability or MSI. This involves copying errors in repeat sequences (Figure 1.4). MSI is most common in colon cancer. It is observed in about 13% of sporadic cases and is the mechanism of cancer initiation in the hereditary non-polyposis colorectal cancer (HNPCC). Another type of small scale instability comes about through defects in nucleotide excision repair genes. These are responsible for the repair of DNA damage caused by exogenous mutagens, most importantly ultraviolet light. It is thus most important in the development of skin cancers. A defect in such repair mechanisms has been found in a disease called *xeroderma pigmentosum*, which is characterized by the development of many skin tumors in sun exposed areas.

Instabilities which involve gross chromosomal alterations are called chromosomal instability or CIN (Figure 1.4). Cells which are characterized by CIN show a variety of chromosomal abnormalities. There can be alterations in chromosome numbers which involve losses and gains of whole chromosomes. This results in aneuploidy. Alternatively, parts of chromosomes may be lost, or we can observe chromosome translocations, gene amplifications, and mitotic recombinations. Many cancers show evidence of chromosomal instability. For example, 87% of sporadic colon cancers show CIN. The reason why CIN is observed in so many cancers is unclear. CIN can be advantageous because it helps to inactivate tumor suppressor genes where both functional copies have to be lost. Assume that one copy of a tumor suppressor gene becomes inactivated by a point mutation which occurs with a rate of 10^{-7} per cell division. The second copy can then be lost much faster by a CIN event (Figure 1.4). For example, CIN could speed up the generation of an APC deficient cell in the colon. On the other hand, CIN is very destructive to the genome. Therefore, even though a cell with an inactivated tumor suppressor gene can be created with a faster rate, clonal expansion of this cell can be compromised because of elevated cell death as a consequence of chromosome loss. The costs and benefits of CIN, as well as the role of CIN in cancer progression, will be discussed extensively in this book.

While it seems intuitive that genetic instability can be advantageous because it leads to the faster accumulation of oncogenic mutations, this is not the whole story. Genetic instability can be advantageous because of an entirely different reason. If cells become damaged frequently, they will enter cell cycle arrest relatively often in order to repair the damage. Therefore, in the presence of elevated damage, repair can compromise the growth of cells. On the other hand, cells which are unstable avoid cell cycle arrest in the face of damage and keep replicating while accumulating genetic alterations. This can lead to an overall higher growth rate of unstable compared to stable cells. The role of DNA damage for the selection of genetic instability will be discussed later in the book.

1.5 Barriers to cancer progression: importance of the microenvironment

So far we have discussed the processes of multi-stage carcinogenesis in some detail. We have thereby concentrated on an approach which is centered around the genetic events which allow cells to escape from growth control and to become cancerous. However, experiments have revealed that the interactions between tumor cells with their tissue micro-environment may be equally important in the process of carcinogenesis [Hsu *et al.* (2002); Tlsty (2001); Tlsty and Hein (2001)]. The stroma surrounding the tumors shows in many cases changes in the patterns of gene expression, in the cellular composition, and in the extracellular matrix. This allows cancers to grow and progress. The development of cancer can thus be seen as a conspiracy between tumor cells and their altered environment which allows uncontrolled growth. Under non-pathogenic conditions, the tissue environment can prevent tumor cells from growing to significant levels.

Interestingly, autopsies have revealed that people who die without ever developing cancers show microscopic colonies of cancer cells which are referred to as *in situ* tumors [Folkman and Kalluri (2004)]. Data suggest that >30% of women in the age range between 40 and 50 who do not develop cancer in their life-time are characterized by small colonies of breast cancer cells. Only 1% of women in this age range, however, develop clinically visible breast cancer. Similar patterns have been observed in the context of thyroid or prostate cancers. The reason for the inability of cancer cells to grow to higher numbers and give rise to pathology is important to understand. The defensive role of the tissue microenvironment in which the cancer tries to grow could be a key factor. For example, cancer cells require the formation of new blood supply in order to obtain oxygen and nutrients, and to grow beyond a relatively small size [Folkman (2002)]. The formation of new blood supply is termed angiogenesis (Figure 1.5).

Our understanding about the role of angiogenesis in the development of

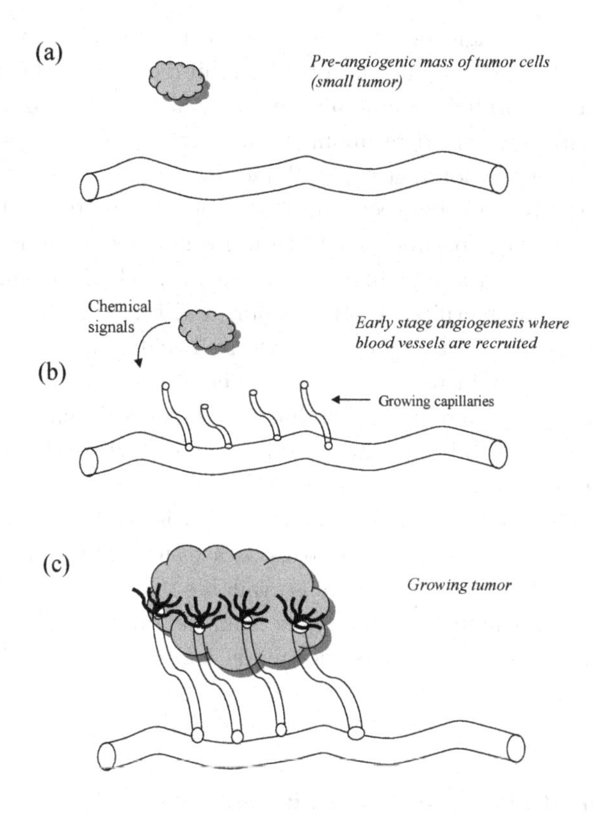

Fig. 1.5 Diagram explaining the concept of angiogenesis. (a) When a cancerous cell is created it can expand up to a small size without the need for blood supply. At this stage, the growth of an *avascular tumor* stops. (b) When angiogenic cell lines emerge, they send out chemical signals called promoters. This induces blood vessels to grow towards the tumor. (c) This process leads to the complete vascularization of the tumor, allowing it to grow to larger sizes.

cancers has been advanced significantly by a variety of studies from Judah Folkman's laboratory [Folkman (1971); Folkman (2002)]. Whether new blood supply can be formed or not appears to be determined by the balance between angiogenesis inhibitors and angiogenesis promoters. Healthy tissue produces angiogenesis inhibitors. Examples of inhibitors are thrombospondin, tumstatin, canstatin, endostatin, angiostatin, and interferons. At the time of cancer initiation, the balance between inhibitors and promoters is heavily in favor of inhibition. Data suggest that even cancer cells themselves initially produce angiogenesis inhibitors which strengthens the defense of the organism against the spread of aberrant genes. In order to grow beyond a small size, angiogenic tumors have to emerge. These are tumor cells which can shift the balance away from inhibition and in favor of promotion. This can be brought about by the inactivation of angiogenesis inhibitors, or by mutations which result in the production of angiogenesis promoters. Examples of promoters are growth factors such as FGF, VEGF, IL-8, or PDGF. If the balance between inhibitors and promoters has been shifted sufficiently in favor of promotion, the cancer cells can grow to higher numbers and progress towards malignancy (Figure 1.5). The mechanisms by which blood supply is recruited to the tumor, and the ways in which inhibitors and promoters affect cancer cells are still under investigation. New blood supply can be built from existing local endothelial cells. On the other hand, angiogenesis promoters may induce a population of circulating endothelial progenitor cells to be recruited to the local site where the blood supply needs to be built. Blood supply can affect cancer cells in two basic ways. First it can influence the rate of cell death. That is, in the absence of blood supply cells die more often by apoptosis as a result of hypoxia, and this is relaxed when sufficient blood supply is available. On the other hand, lack of blood supply can prevent cancer cells from dividing. In this case they remain *dormant*, That is, they do no divide and do not die. These dynamics will be discussed extensively.

1.6 Evolutionary theory and Darwinian selection

Theodosius Dobzhansky who, according to Stephen J. Gould, was the greatest evolutionary geneticist of our times, wrote that "nothing in biology makes sense except in the light of evolution". This also applies to our understanding of cancer. The process of carcinogenesis includes all the essential ingredients of evolutionary theory: reproduction, mutation, and selection (Figure 1.6).

As outlined in detail above, the entire process of cancer initiation and progression is concerned with the accumulation of mutations which allow the cells to break out of normal regulatory mechanisms. Such cells will grow better than healthy cells and are advantageous. In evolutionary terms, they are said to have a higher *fitness*. The more oncogenic mutations the cells acquire, the better they are adapted to growing in their environments, and the higher their fitness. Cancer cells which grow best can be selected for and can exclude less fit genotypes. Cancer cells can even adapt their "evolvability": genetically unstable cells may be able to evolve faster and adapt

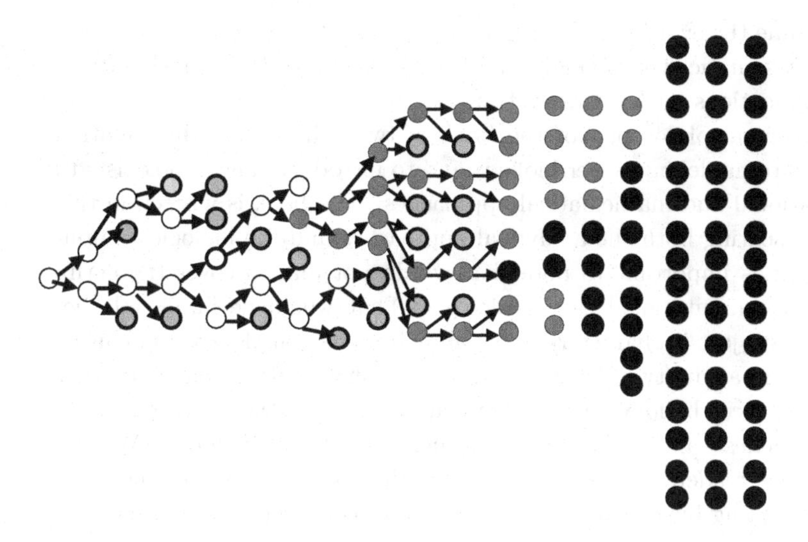

Fig. 1.6 Diagram explaining the concept of somatic evolution and cancer progression. Cancer originates with the generation of a mutant cell. This cell divides and the population grows. This is called clonal expansion. Further mutations can subsequently arise which have a higher fitness. They grow and expand further. Consecutive mutations and rounds of clonal expansion allow the cancer to grow to ever increasing sizes.

better than stable cells. This can be very important in the face of many selective barriers and changing environments. Barriers can include inhibitory effects which are exerted by the tissue microenvironment, or an adaptive immune system which can specifically recognize a variety of tumor proteins and mount new responses as the tumor evolves. The environment in our bodies can change over time and render different genotypes advantageous at different stages. An example is aging which involves the continuous rise in the rate of DNA damage as a result reactive oxygen species which are produced as a byproduct of metabolism.

The somatic evolution of cells will be a central component of the mathematical models which are discussed in this book. A large part of the chapters will investigate the selective forces which can account for the emergence of genetic instability in cancer. Is instability selected for because it allows faster adaptive evolution of cells as a result of the enhanced ability to acquire oncogenic mutations? Can instability reduce the fitness of cells because it destroys the integrity of the genome? Can a rise in the level of DNA damage select for unstable cells because the costs of arrest and senescence are avoided? What are the pathways to cancer? Further, the book will examine the evolution of angiogenic cell lines, the relationship between immunity, somatic evolution and cancer progression, and will conclude with some implications for treatment strategies.

The central philosophy of the book is twofold: to introduce mathematicians to modeling cancer biology, and to introduce cancer biologists to computational and mathematical approaches. The book is written in this spirit, presenting both the analytical approaches and the biological implications. It is important to note that we do not aim to cover the entire subject of computational cancer biology. That would be impossible because the subject is characterized by an enormous complexity and can be addressed on a variety of levels. Instead, we concentrate on one particular aspect of cancer biology; that is, we concentrate on studying the process of carcinogenesis and consider it in the light of somatic evolution. We aim to introduce readers to the basic mathematical methodology as well as to some interesting biological insights which have come out of this work.

Chapter 2

Mathematical modeling of tumorigenesis

Traditionally, scientists from many different backgrounds have been producing interesting and unexpected ideas. Their methods come from various fields, including applied mathematics, statistics, computer science, material science, fluid mechanics, population dynamics and evolutionary game theory.

Broadly speaking, there are three major areas where theory has contributed the most to cancer research:

(i) **Modeling in the context of epidemiology and other statistical data.** One of the oldest and most successful methodologies in theoretical cancer research is using the available incident statistics and creating models to explain the observations. This field was originated by Armitage and Doll in 1954 [Armitage and Doll (1954)], and then taken to the next level by Moolgavkar and colleagues [Moolgavkar and Knudson (1981)].

(ii) **Mechanistic modeling of avascular and vascular tumor growth.** An entirely different approach to cancer modeling is to look at the mechanistic aspects of tumor material and use physical properties of biological tissues to describe tumor growth, see [Preziosi (2003)] for review.

(iii) **Modeling of cancer initiation and progression as somatic evolution.** In this area of research, methods of population dynamics and evolutionary game theory are applied to study cancer. First developed by ecologists and evolutionary biologists, these methods have been used to understand the collective behavior of a population of cancer cells, see [Gatenby and Gawlinski (2003)], [Gatenby and Vincent (2003b)], [Gatenby and Vincent (2003a)].

In this Chapter we review basic mathematical tools necessary to undertake different types of cancer modeling. These are: ordinary differential equations, partial differential equations, stochastic processes, cellular automata and agent based modeling.

2.1 Ordinary differential equations

Mathematical modeling of growth, differentiation and mutations of cells in tumors is one of the oldest and best developed topics in biomathematics. It involves modeling of growth, differentiation and mutation of cells in tumors. Let us view cancer as a population of cells, which has some potential to grow. In the simplest case, we can model cellular growth followed by saturation with the following logistic ordinary differential equation (ODE):

$$\dot{x} = rx(1 - x/k), \quad x(0) = 1,$$

where dot is the time derivative, $x = x(t)$ is the number of cancer cells at time t, r is the growth rate and k is the carrying capacity, that is, the maximal size the population of cells can reach, defined by the nutrient supply, spatial constraints etc. The solution of the above ODE is a familiar looking "sigmoidal" curve.

Next, let us suppose that the population of cells is heterogeneous, and all cells compete with each other and with surrounding healthy cells for nutrients, oxygen and space. Then we can imagine the following system, equipped with the appropriate number of initial conditions:

$$\dot{x}_i = r_i x_i - \phi x_i, \quad 0 \leq i \leq n, \quad x_i(0) = \hat{x}_i,$$

where x_i is the number of cells of type i, with the corresponding growth rate, r_i. We have the total of n types, and we can model the competition by the term ϕ in a variety of ways, e.g. by setting

$$\phi = \frac{\sum_{i=0}^{n} r_i x_i}{N},$$

where $N = \sum_{i=0}^{n} \hat{x}_i$ is the total number of cells in the system, which is assumed to be constant in this model.

As a next step, we can allow for mutations in the system. In other words, each cell division (happening with rate r_i for each type) has a chance to result in the production of a different type. Let us assume for simplicity that the type i can mutate into type $(i+1)$ only, according to the following simple diagram:

$$x_0 \to x_1 \to \ldots \to x_{n-1} \to x_n$$

Then the equations become,

$$\begin{aligned}
\dot{x}_0 &= r_0(1 - u_0)x_0 - \phi x_0, \\
\dot{x}_i &= u_{i-1}r_{i-1}x_{i-1} + r_i(1 - u_i)x_i - \phi x_i, \quad 1 \leq i \leq n-1, \\
\dot{x}_n &= r_{n-1}u_{n-1}x_{n-1} + r_n x_n - \phi x_n, \\
x_i(0) &= \hat{x}_i, \quad 0 \leq i \leq n,
\end{aligned}$$

where ϕ is defined as before, and u_i is the probability that a cell of type $(i + 1)$ is created as a result of a division of a cell of type i. The above equations are called the *quasispecies* equations. These were introduced by Manfred Eigen in 1971 as a way to model the evolutionary dynamics of single-stranded RNA molecules in *in vitro* evolution experiments. Since Eigen's original paper, the quasispecies model has been extended to viruses, bacteria, and even to simple models of the immune system. Quasispecies equations are *nonlinear*, like most differential equations in cancer modeling. However, there is a simple and elegant way to solve these equations, which we review in Chapter 7. In a more general case, the mutation network can be more complicated, allowing mutations from each type to any other type. This is done by introducing a mutation *matrix* with entries, u_{ij}, for mutation rates from type i to type j. Examples of using quasispecies equations in recent literature are [Sole and Deisboeck (2004)] and [Gatenby and Vincent (2003b)].

Other ordinary differential equations used to study the dynamics of cancerous cells are similar to predator-prey systems in ecology. For instance, Gatenby and Vincent [Gatenby and Vincent (2003a)] used the following competition model,

$$\dot{x} = r_x \left(1 - \frac{x + \alpha_{xy}y}{k}\right)x, \quad \dot{y} = r_y \left(1 - \frac{y + \alpha_{yx}x}{k}\right)y,$$

where x and y describe the populations of cancerous and healthy cells, respectively. Moore and Li [Moore and Li (2004)] used a model in a similar spirit to describe the interactions between chronic myelogenous leukemia (CML) and T cells. They considered a system of 3 ODEs, for naive T cells, effector T cells and CML cancer cells.

The equations shown above are toy models to illustrate general principles, rather than actual tools to study real biological phenomena. However, by modifying these equations and incorporating particular properties of a biological system in question, we can describe certain aspects of cancer dynamics. Like any other method, the method of ODEs has advantages and drawbacks. Among the advantages is its simplicity. The disadvantages include the absence of detail. For instance, no spatial interactions can be described by ODEs, thus imposing the assumption of "mass-action"-type interactions. Stochastic effects are not included, restricting the applicability to large systems with no "extinction" effects.

Finally, because of an empirical nature of this kind of modeling, this method (like most other empirical methods) presents a problem when trying to find ways to *measure* coefficients in the equations. Several methods of *robustness* analysis have been developed. The main idea is as follows. If the number of equations is in the tens, and the number of coefficients is in the hundreds, one could argue that almost any kind of behavior can be reproduced if we tune the parameters in the right way. Therefore, it appears desirable to reduce the number of unknown parameters and also to design some sort of reliability measure of the system. In the paper by [Moore and Li (2004)], Latin hypercube sampling on large ranges of the parameters is employed, which is a method for systems with large uncertainties in parameters. This involves choosing parameters randomly from a range and solving the resulting system numerically, trying to identify the parameters to which the behavior is the most sensitive. In the paper by Evans et al [Evans *et al.* (2004)], "structural identifiability analysis" is discussed, which determines whether model outputs can uniquely determine all of the unknown parameters. This is related to (but is not the same as) the confidence with which we view parameter estimation from experimental data. In general, questions of robustness and reliability are studied in *mathematical control theory.*

2.2 Partial differential equations

The next method that we will mention here is partial differential equations (PDEs). They can be a very useful tool when studying tumor growth and invasion into surrounding tissue. In many models, tumor tissue is described as a mechanistic system, for instance, as a fluid (with a production term proportional to the concentration of nutrients) [Evans *et al.* (2004)], or as a mixture of solid (tumor) and liquid (extracellular fluid with nutrients) phases [Byrne and Preziosi (2003)]. As an example, we quote the system used by Franks et al. [Fauth and Speicher (2001); Franks *et al.* (2003)]. These authors view an *avascular* tumor as a coherent mass whose behavior is similar to that of a viscous fluid. The variables $n(\mathbf{x}, t)$, $m(\mathbf{x}, t)$ and $\rho(\mathbf{x}, t)$ describe the concentration of tumor cells, dead cells and surrounding material, respectively. The nutrient concentration is $c(\mathbf{x}, t)$, and the velocity of cells is denoted by $\mathbf{v}(\mathbf{x}, t)$. Applying the principle of mass balance to different kinds of material, we arrive at the following system:

$$\dot{n} + \nabla \cdot (n\mathbf{v}) = (k_m(c) - k_d(c))n, \tag{2.1}$$

$$\dot{m} + \nabla \cdot (m\mathbf{v}) = k_d(c)n, \tag{2.2}$$

$$\dot{\rho} + \nabla(\rho\mathbf{v}) = 0. \tag{2.3}$$

Here, we have production terms given by the rate of mitosis, $k_d(c)$, and cell death, $k_d(c)$, which are both given empirical functions of nutrient concentration. The nutrients are governed by a similar mass transport equation,

$$\dot{c} + \nabla(c\mathbf{v}) = D\nabla^2 c - \gamma k_m(c)n,$$

where D is the diffusion coefficient and $\gamma k_m(c)n$ represents the rate of nutrient consumption. In order to fully define the system, we also need to use the mass conservation law for the cells, modeled as incompressible, continuous fluid, $n + m + \rho = 1$. Finally, a constitutive law for material deformation must be added to define the relation between concentration (stress) and velocity. Also, the complete set of boundary conditions must be imposed to make the system well defined. We skip the details here, referring the reader to the original papers. Our goal in this chapter is to give the flavor of the method.

As our next example, we will mention the paper by Owen et al. [Owen *et al.* (2004)] which modeled the use of macrophages as vehicles for drug delivery to hypoxic tumors. This model is based on a growing avascular tumor spheroid, in which volume is filled by tumor cells, macrophages and extracellular material, and tumor cell proliferation and death is regulated by nutrient diffusion. It also includes terms representing the induced death of tumor cells by hypoxic engineered macrophages. Matzavinos et al. [Matzavinos *et al.* (2004)] used four nonlinear PDEs for tumor-infiltrating cytotoxic lymphocytes, tumor cells, complexes and chemokines to describe the interaction of an early, prevascular tumor with the immune system.

Avascular growth is relevant only when studying very small lesions, or tumor spheroids grown *in vitro*. To describe realistically tumorigenesis at later stages, one needs to look at the vascular stage and consider mechanisms responsible for angiogenesis. An extension of avascular mechanistic models which includes angiogenesis can be found for example in the paper by Breward et al. [Breward *et al.* (2003)]. There, cells divide and migrate in response to stress and lack of nutrition. Here we present a different model developed by Anderson and Chaplain [Anderson and Chaplain (1997)]. It describes the dynamics of endothelial cell (EC) density, migrating toward a tumor and forming neovasculature in response to specific chemical signals, *tumor angiogenic factors* (TAF). If we denote by $n(\mathbf{x}, t)$ the EC density, then their migration can be described as

$$\dot{n} = D\nabla^2 n - \nabla(\chi(c)n\nabla c) + g(n, f, c),$$

where D and $\chi(c)$ are the diffusion and the chemotactic parameter respectively, $c(\mathbf{x}, t)$ is a specific chemical (TAF) responsible for chemotaxis, and $g(n, c)$ is the proliferation function. In the simplest case, the chemicals can be assumed to be in a steady-state (that is, t-independent), or they can satisfy a PDE:

$$\dot{c} = D_c\nabla^2 n + v(c, n),$$

with $v(n, f)$ being a specific production/uptake function.

As with any system of nonlinear PDEs, one should be careful about well-posedness of the problem. The appropriate boundary conditions must be imposed, depending on the dimensionality and geometry of the problem. Then, either numerical solutions can be investigated, or a stability analysis of a simple solution performed. An example of a simple solution could be, for instance, a spherically symmetrical or planar tumor.

These could be interesting exercises in applied mathematics, but in our experience, description of most biologically relevant phenomena does not readily follow from a stability analysis and requires a specific, directed question. Mechanistic models are necessarily an idealization of reality, and the only way to judge how "bad" such a description is comes from formulating a specific biological question and figuring out what is necessary and what is secondary with respect to the phenomenon in question.

As an example of an interesting usage of the mechanistic tumor modeling by PDEs, where experience of physical sciences is used to study specific biological phenomena, we will mention the paper on the phenomenon of vascular collapse [Araujo and McElwain (2004)]. In this study, the distribution of stress is calculated throughout the tumor, as it changes in time as a result of cell division. The vascular collapse is modeled by identifying the region where stresses exceed a critical value. At this point, a collapse occurs and the inner regions of the tumor become cut out of the central blood system. The growth of tumor obtained in the model resembles experimentally observed patterns, with exponential growth before and retardation after the collapse. Another paper that we would like to mention addressed the question of the precise origin of neovascularization [Stoll *et al.* (2003)]. The traditional view of angiogenesis emphasizes proliferation and migration of local endothelial cells. However, circulating endothelial progenitor cells have recently been shown to contribute to tumor angiogenesis. The paper quantified the relative contributions of endothelial and endothelial progenitor cells to angiogenesis using a mathematical model, with implications for the rational design of antiangiogenic therapy.

At the next level of complexity, we have integro-differential equations. They can be used to describe nonlocal effects or inhomogeneity of the population of cells, such as age structure. For an example of integro-differential equations in tumor modeling, see [Dallon and Sherratt (1998)].

To summarize the method of partial differential equations, applied to mechanistic modeling of tumor growth, we note that it is significantly more powerful than the method of ODEs, as it allows us to make a dynamic description of spatial variations in the system. We have a large, well-established apparatus of mathematical physics, fluid mechanics and material science working for us, as long as we model biological tissue as a "material". We have the comforting convenience of *laws* as long as we are willing to stay within the realm of physics (complex biological systems cannot boast having any laws whatsoever). The problem is that we do not exactly know to what extent a tumor behaves as an incompressible fluid (or homogeneous porous medium, or any other physical idealization), and to what extent its behavior is governed by those mysterious biological mechanisms that we cannot fit into a neat theory. Any researcher using the framework of mechanistic modeling should be prepared to adapt the model to allow for more biology.

On a more down-to-earth note, there is one obvious limitation of PDEs which comes from the very nature of differential equations: they describe continuous functions. If the cellular structure of an organ is important, then we need to use a different method, and this is what we consider next.

2.3 Discrete, cellular automaton models

Cellular automaton models are based on a spatial grid, where the dynamics is defined by some local rules of interaction among neighboring nodes. The interaction rules can be deterministic or stochastic (that is, dictated by some random processes, with probabilities imposed). Each grid point may represent an individual cell, or a cluster of cells; for simplicity we will refer to them as "cells". To begin, we present a very simple model of tumor growth which illustrates the method. We start from a rectangular, two-dimensional grid. Let us refer to a grid point as x_{ij}, where i and j are the horizontal and vertical coordinates of the point. Each node, x_{ij}, can be a healthy cell, a cancer cell, or a dead cell. We start from an initial distribution of tumor cells, healthy cells and dead cells. For each time-step, we update the grid values according to some local interaction rules. Let us denote the discrete time variable as $n = 1, 2, \ldots$. Here is an example of an update rule. At each time step, we update one site, in a fixed order; $x_{ij}(n+1)$ is given by the following:

- If x_{ij} is surrounded by a layer of thickness δ of tumor cells, it dies.
- If x_{ij} is a tumor cell not surrounded by a δ-layer of tumor cells, it reproduces. This means that the site x_{ij} remains a tumor cell, and in addition, one of its neighbors (chosen at random) becomes a tumor cell. As a result, all the non-dead cells on that side of x_{ij} between x_{ij} and the nearest dead cell, are shifted away from x_{ij} by one position.
- If x_{ij} is a dead cell, it remains a dead cell.
- If x_{ij} is a healthy cell, it remains a healthy cell.

Of course, this is only a toy model. Cellular automata models used to describe realistic situations are more complicated as they have to grasp many aspects of tumor biology. For instance, Kansal et al. [Kansal *et al.* (2000)] used a very sophisticated three-dimensional cellular automaton model of brain tumor growth. It included both proliferative and non-proliferative cells, an isotropic lattice, and an adaptive grid lattice.

Cellular automata have been used to study a variety of questions. Alarcon et al. [Alarcon *et al.* (2003)] studied how inhomogeneous environments can affect tumor growth. They considered a network of normal healthy blood vessels and used an (engineering in spirit) approach to model the dynamics of blood flow through this fixed network. The outcome of this part of the model was the distribution of oxygen (red blood cells) throughout the network. Next, a cellular automaton model was run where, like in our toy model above, each element of the discrete spatial grid could take one of three values: "unoccupied", "has a normal cell", or "has a cancerous cell". The concentration of oxygen was fed into the local interaction rules.

In the paper by Gatenby and Gawlinski [Gatenby and Gawlinski (2003)], the acid-mediated tumor invasion hypothesis was studied. This hypothesis states that tumors are invasive because they perturb the environment in such a way that it is optimal for their proliferation, and toxic to the normal cells with which they compete for space and substrate. The authors considered a spatial tumor invasion model, using PDEs and cellular automata. The model was based on the competition of healthy and tumor cells, with elements of acid production by tumor cells, acid reabsorption, buffering and spatial diffusion of acid and cells. The authors propose that the associated *glycolytic phenotype* represents a successful adaptation to environmental selection parameters because it confers the ability for the tumor to invade.

A cellular automaton model of tumor angiogenesis was designed by [Anderson and Chaplain (1998)]. In their discrete model, the movement rules between states are based directly on a discretized form of the continuous model, which was considered in the previous section. The discretization is performed by using the Euler finite difference approximation to the PDEs. Then, numerical simulations allow for tracking the dynamics of individual endothelial cells, as they build blood vessels in response to TAFs. A qualitatively novel feature of this model is its ability to describe branching of new vessels by imposing some simple local rules. In particular, it is assumed that if the density of TAFs is above critical, there is enough space for branching, and the current sprout is sufficiently "old", then there is a finite probability for the vessel to branch and form a new sprout. This behavior cannot be grasped by the continuous, PDE-based models.

Finally, we would like to mention that apart from cellular automaton type models, there is an emerging area of agent-based modeling applied to tumor growth. In such models, each cell is modeled as an agent with some "strategy". In a paper by [Mansury and Deisboeck (2003)], an agent-based model was applied to investigate migration of tumor cells. It was assumed that tumor cells can proliferate or migrate, depending on nutrition and toxicity of their environment, using a local search algorithm. It turned out that the search precision did not have to be 100% to ensure the maximum invasion velocity. It had a saturation level, after which it did not pay to overexpress the genes involved in the search.

To conclude the section on discrete modeling, we would like to mention that the cellular automaton approach gives rise to a new class of behaviors which can hardly be seen in continuous, PDE-based models. It allows to track individual cells, and reproduce the dynamics of emerging structures such as tumor vasculature. A drawback of this approach is that it is almost universally numerical. It is difficult to perform any analysis of such models, which leaves the researcher without an ability to generalize the behavioral trends.

2.4 Stochastic modeling

Next we review one of the most important tools in biological modeling, which is stochastic processes. The need for stochastic modeling arises because many of the phenomena in biology have characteristics of random variables. That is, as a process develops, we cannot predict exactly the state of the system at any given moment of time, but there are certain trends which can be deduced, and, if repeated, an experiment will lead to a similar in some sense (but not identical) outcome. In this chapter we are not aiming at a comprehensive introduction to stochastic processes. Rather, we will give several examples where various stochastic methods are used to describe tumorigenesis.

The process where the stochastic nature of events can be seen very clearly is the accumulation of mutations. This process is central to cancer progression, and therefore developing tools to describe it is of vital importance for modeling. In the simplest case, we can envisage cell division as a binary (or *branching*) process, where at regular instances of time, each cell divides into two identical cells with probability $1 - u$, and it results in creating one mutant and one wild type cell with probability u. To complete the description of this simplified model, we assume that a mutant cell can only give rise to two mutant daughter cells. Let us start from one wild type cell and denote the number of mutants at time n as z_n. The random variable z_n can take nonnegative integer values; another way to say this is that the state space is $\{0\} \cup I$. This is a simple branching process, which is a discrete state space, discrete time process. We could ask the question: what is the probability distribution of the variable z_n? Possible modifications of this process can come from the existence of several consecutive mutations, a possibility of having one or both daughter cells mutate as a result of cell division, allowing for cell death, or from distinguishing among different kinds of mutations. As an example of a branching process type model, we will mention the recent paper by Frank [Frank (2003)] which addressed the accumulation of somatic mutation during the embryonic (developmental) stage, where cells divide in a binary fashion, similar to the branching process. Two recessive mutations to the retinoblastoma locus are required to initiate tumors. In this paper, a mathematical framework is developed for somatic mosaicism in which two recessive mutations cause cancer. The following question is asked: given the observed frequency of cells with two mutations, what is the conditional frequency distribution of cells carrying one mutation (thus rendering them susceptible to transformation by a second mutation)? *Luria-Delbruck*-type analysis is used to calculate a conditional distribution of single somatic mutations.

Next, we consider another important process, the *birth and death process*. Suppose that we have a population of cells, whose number changes from time t to time $t + \Delta t$, where Δt is a short time interval, according to the following rules:

- With probability $L\Delta t$ a cell reproduces, creating an identical copy of itself,
- With probability $D\Delta t$ a cell dies.

All other events have a vanishingly small probability. The number of cells, $X(t)$, can take positive integer values, and it depends on the continuous time variable. That is, it can change at any time, and not just at prescribed intervals. Therefore, this is a continuous time, discrete state space process. One obvious modification to the above rules is to include mutations. Say, instead of $L\Delta t$, we could have the probability $L(1 - u)\Delta t$ to reproduce faithfully, and probability $Lu\Delta t$ to create a mutant. Further, we could consider a chain of mutations, and describe the evolution of the number of cells of each type. This resembles Moolgavkar's description of multistage carcinogenesis [Moolgavkar and Knudson (1981)] which is reviewed in Chapter 3.

In the birth-death type processes, the population of cells may become extinct, or it could grow indefinitely. Another type of process that is very common in tumor modeling corresponds to constant population size. An example is the Moran process. Whenever a cell reproduces (with the probability weighted with the cell's fitness), another cell is chosen to die to keep a constant population size. If we include a possibility of mutations (or sequences of mutations), which lead to a change of fitness in cells, we can model an emergence and invasion of malignant cells. Models of this kind are relevant for the description of cellular compartments [Komarova *et al.* (2003)] or organs of adult organisms. In a series of stochastic models, Frank and Nowak [Frank and Nowak (2003)] discussed how the architecture of renewing epithelial tissues could affect the accumulation of mutations. They showed that a hierarchy of stem cells could reduce the accumulation of mutations by the mechanism that they term *stochastic flushing.* They assume that each compartment retains a pool of nearly quiescent proto-stem cells. The renewal of tissue happens in the usual way by stem cell divisions. If a stem cell dies, it is replaced from the pool of proto-stem cells. This process is characterized by the absence of long stem cell lineages, which protects tissue from accumulating mutations [Michor *et al.* (2003b)]. The interesting question of the role of compartment size on the accumulation of somatic mutations in cancer was addressed by [Michor *et al.* (2003a)]. They assumed that the total number of cells in the organ is fixed, and divided the population into compartments of variable size, N. Then they used a Moran process to calculate the optimum value, N, which minimizes the rate of accumulation of mutant cells.

Stochastic models of stem cell dynamics have been proposed by many authors. [Nowak *et al.* (2003)] employ a *linear process* of somatic evolution to mimic the dynamics of tissue renewal. There, cells in a constant population are thought to be put in a straight line. The first one is the symmetrically dividing stem cell, which places its offspring next to itself and moves the other cells by one position. The last cell is taken out of the system. This process has the property of canceling out selective differences among cells yet retaining the protective function of apoptosis. It is shown that this design can slow down the rate of somatic evolution and therefore delay the onset of cancer. A different constant population model is employed by [Calabrese *et al.* (2004); Kim *et al.* (2004)], where precancerous mutations in colon stem cell compartments (*niches*) are studied. Each niche contains multiple stem cells, and niche stem cells are lost at random with replacement. It is assumed that each stem cell can either divide asymmetrically, or give rise to two stem cells, or to two differentiated cells. This loss and replacement dynamics eventually leads to the loss of all stem cell lineages except one. The average time to cancer is calculated with this model, using five successive mutational steps. The results are compared with the existing age-incidence statistics. We will briefly mention statistical methods in the next subsection.

2.5 Statistics and parameter fitting

The idea is the following. A multi-stage model of carcinogenesis is formulated as a stochastic process, which includes a series of mutational events and clonal expansions. The mutation rates, the average rates of clonal expansions for each stage, and even the number of stages are variables of the model. Then, the probability of developing cancer by a certain age is calculated (usually, by means of numerical simulations), as a function of all the unknown parameters. The outcome of such calculations, for each set of parameters, is then compared with the existing data on cancer incidence, and the set of parameters which gives the best fit is identified. In their excellent paper, Luebeck and Moolgavkar [Luebeck and Moolgavkar (2002)] use the data on the incidence of colorectal cancers in the Surveillance, Epidemiology, and End Results (SEER) registry. They conclude that the statistics are most consistent with a model with two rare events followed by a high-frequency event in the conversion of a normal stem cell into an initiated cell that expands clonally, which is followed by one more rare event. The two rare events involved in the initiation are interpreted to represent the homozygous loss of the APC gene.

Many authors have analyzed age-incidence curves [Frank (2004); Haylock and Muirhead (2004); Krewski *et al.* (2003)] and death statistics [Filoche and Schwartz (2004)]. In thc latter paper, the statistics of fluctuations in cancer deaths per year lead to an intriguing discovery: there is a big difference between cancers of young ages and cancers after 40. The authors suggest that cancers attacking older people behave like "critical systems" in physics and can be considered as an avalanche of "malfunctions" in the entire organism.

Another interesting way of looking at cancer statistics (of a different kind) is due to Mitelman and his colleagues. They study distributions of chromosome aberrations in various cancers and use statistical tools to analyze the emerging patterns [Hoglund *et al.* (2002a); Hoglund *et al.* (2002b); Hoglund *et al.* (2001); Mitelman (2000)]. In one paper [Frigyesi *et al.* (2003)], they found that the number of chromosomal imbalances per tumor follows a power law (with the exponent one). The main idea of a model explaining this behavior is as follows. The karyotype in unstable cancers evolves gradually, in such a way that the variability is proportional to the number of changes that already exist. The authors propose two possible interpretations of the model. One is that the rate at which changes accumulate, increases as cancer progresses (this is consistent with the notion of a "mutator phenotype" due to Loeb [Loeb (2001)]). The other is the evolving and increasingly permissive tumor environment. Similar theoretical and computational tools were applied to testicular germ cell tumor karyotypes [Frigyesi *et al.* (2004)]. It was shown that two distinct processes are operating in the karyotypic evolution of these tumors; whole-chromosome changes originate from a multipolar cell division of a tetraploid cell, whereas imbalances accumulate in a stepwise manner.

2.6 Concluding remarks

In the rest of the book, we will employ many of the methods mentioned here, and explain models and their solutions in detail. In particular, Chapter 3 talks about birth-death processes; Chapter 5 uses a branching process; the Moran process is employed in Chapters 3 and 4; Chapter 6 combines a stochastic Moran process with a deterministic ODE; Chapter 7 studies quasispecies ODEs; Chapter 8 uses PDEs and pattern formation-type analysis; and Chapters 9–12 talk about analytical and numerical solutions of ODEs.

Chapter 3

Cancer initiation: one-hit and two-hit stochastic models

The question of the origins of cancer is among the most important in our understanding of the disease. There is no universal answer to this question, as different cancers are initiated by different mechanisms. There are however certain patterns that can be recognized. Among the most important ones is cancer initiation via the inactivation of a tumor suppressor gene. The concept has evolved during the last 30 years. A defining landmark was the discovery of the Rb gene.

Retinoblastoma is a rare and deadly cancer of the eye that afflicts children. It comes in two versions. One affects newborn infants and is characterized by multiple tumors. The other hits children when they are older and is usually characterized by only a single tumor. In 1971, Alfred Knudson proposed an explanation, which became known as the famous Knudson's "two-hit hypothesis" [Knudson (1971)]. According to his theory, in the early-onset version of retinoblastoma, children inherit a defective gene from one parent. These children are halfway to getting the disease the moment they are born. Then, an error in DNA replication in a single eye cell, causing a defect in the normal gene that was inherited from the other parent, would send that cell on its way to becoming a tumor. In contrast, children who develop retinoblastoma later in childhood are probably born with two good copies of the gene but acquire two hits in both copies of the gene in a cell. This would take longer, causing the cancer to show up at a later age.

Knudson proposed the tumor suppressor gene hypothesis of oncogenesis after detecting a partial deletion of chromosome 13 in a child with retinoblastoma. This was a revolutionary concept, that is, cancer was not caused by the *presence* of an oncogene, but rather the *absence* of an "anti-oncogene." He concluded that the retinoblastoma tumor suppressor gene would be found at band 13q14. It wasn't until the late 1980s when scientists eventually cloned the gene Rb which mapped exactly to the location predicted by Knudson.

Other genes with similar properties were discovered, including p53, WT1, BRCA1, BRCA2 and APC. The generic definition of a *tumor suppressor gene* comprises the idea of a *loss of function*. Only when both alleles of the gene are inactivated, does the cell acquire a phenotypic change. Many of tumor suppressor genes are involved in familial cancers. The mechanism is similar to the one described by Knudson in retinoblastoma. If a defective allele is present in the germline, the affected individuals will have a higher chance of developing a cancer as only one remaining allele must be inactivated to initiate an early stage lesion.

In collaboration with Knudson, Suresh Moolgavkar went on to develop mathematical models for this hypothesis, which were the first to coalesce clinical-epidemiological observations with putative mutation rates and molecular genetics [Moolgavkar and Knudson (1981)]. In subsequent publications, Moolgavkar and colleagues have created a rigorous methodology of studying multistage carcinogenesis [Moolgavkar (1978); Moolgavkar *et al.* (1980); Moolgavkar *et al.* (1988)]. In this chapter we will review some of the main ideas of the two-hit models, and develop them further to provide tools for this book. In particular, we will derive simple expressions for the probability of generating double-mutants. We consider small, intermediate and large populations, in the case of disadvantageous, neutral or advantageous intermediate mutants. We start from a *one-hit model* and then go on to describe a more involved process with two hits.

3.1 A one-hit model

We will use this section to review several important mathematical tools describing stochastic population dynamics.

3.1.1 *Mutation-selection diagrams and the formulation of a stochastic process*

Let us first assume that there are two types of cells in a population, which we will call type "A" and type "B". Cells can reproduce, mutate and die. The probability that a cell of type "A" reproduces faithfully is $1 - u$; with probability u it will mutate to type "B". Cells of type "B" always reproduce faithfully. We will assume that the total number of cells is constant and equal to N. Let the cells of type "A" have reproductive rate 1 and the cells of type "B" - reproductive rate r.

We will use the following convenient short-hand representation of these processes:

$$A_{(1)} \xrightarrow{\quad u \quad} B_{(r)} \tag{3.1}$$

Here the reproductive rate of each type is given in brackets and the mutation rate is marked above the arrow. We will refer to such diagrams as *mutation-selection networks*.

The one-hit model can be relevant for the description of an oncogene activation, or cancer initiation in patients with familial disorders, where the first allele is mutated in the germ line, and the inactivation of the second allele leads to a fitness advantage of the cell. In these cases, we can assume $r > 1$. In the more general case, we can view the one-hit model as the process of any one genetic alteration, resulting in a advantageous ($r > 1$), disadvantageous ($r < 1$) or a neutral ($r = 1$) mutant.

The Moran process. One can envisage the following birth-death process (called the *Moran process*). At each time step, one cell reproduces, and one cell dies. We set the length of each time step to be $1/N$, so that during a unit time interval, N cells are chosen for reproduction and N cells die. We assume that all cells have an equal chance to die (this is equal to $1/N$). On the other hand, reproduction happens differentially depending on the type, and the relative probability of being chosen for reproduction is given by 1 and r for the cells of types "A" and "B" respectively. Obviously, in this setting the total number of cells is preserved.

Let us denote the number of cells of type "A" as a, and the number of cells of type "B" as b, so that $a + b = N$. The probability that a cell of type "A" reproduces is proportional to its frequency and the reproductive rate, and is given by $a/(a + rb)$. Similarly, the probability that a cell of type "B" reproduces is $rb/(a + rb)$. Thus the probability that the new cell is of type "A" or type "B" is given respectively by

$$P_{+A} = (1-u)\frac{a}{a+rb}, \quad P_{+B} = u\frac{a}{a+rb} + \frac{rb}{a+rb}.$$

Cells of both types have a probability to die proportional to their abundance, i.e. the probability that a cell of type "A" (or "B") dies is given respectively by

$$P_{-A} = \frac{a}{N}, \quad P_{-B} = \frac{b}{N}.$$

We will refer to an event consisting of one replication and one cell death by an *elementary event.*

The resulting population dynamics is a Markov process with states $b = 0, 1, \ldots, N$, and time steps of length $1/N$. The probability that an elementary event results in an increase of the number of cells of type "B", is equal to $P_{+B}P_{-A}$, and the probability that the number of cells of type "B" decreases is equal to $P_{-B}P_{+A}$. If P_{ij} is the probability to go to state $b = j$ from state $b = i$, then the transition matrix is given by

$$P_{ij} = \begin{cases} \frac{u(N-i)+ri}{\mathcal{N}_i}\frac{N-i}{N}, & j = i+1, \\ \frac{(1-u)(N-i)}{\mathcal{N}_i}\frac{i}{N}, & j = i-1, \\ 1 - P_{i,i+1} - P_{i,i-1} & j = i, \\ 0 & \text{otherwise}, \end{cases} \tag{3.2}$$

where $0 \le i, j \le N$, and we introduced the notation

$$\mathcal{N}_i = N - i + ir.$$

The corresponding Markov process is a biased random walk with one absorbing state, $b = N$. Let us set the initial condition to be $b = 0$ (all cells are of type "A") and study the dynamics of absorption into the state $b = N$.

Notation for the time-variable. In this chapter, we will adopt the upper case variable, T, for measuring the time in terms of elementary events. The lower-case notation is reserved for time measured in terms of *generations*. For constant population processes, we have the simple relation,

$$t = T/N.$$

3.1.2 *Analysis of a one-hit process*

Diffusion approximation. Let us denote the probability to be in state $a = i$ at time T as $\varphi_i(T)$. Using the transition matrix for two types, (3.2), we can write down the Kolmogorov forward equation for φ:

$$\frac{\partial \varphi_i}{\partial T} = (1-u)\left[\varphi_{i-1}\frac{(i-1)[N-(i-1)]}{\mathcal{N}_{i-1}} - \varphi_i \frac{i(N-i)}{\mathcal{N}_i}\right] + \varphi_{i+1}\frac{[r(N-(i+1))+(i+1)u](i+1)}{\mathcal{N}_{i+1}} - \varphi_i \frac{[r(N-i)+iu](N-i)}{\mathcal{N}_i}.$$

It is convenient to introduce the variable $\eta = i/N$. Taking the continuous limit and expanding into the Taylor series up to the second order, we obtain the following partial differential equation for $\varphi(\eta, T)$:

$$\frac{\partial \varphi}{\partial T} = \frac{\partial}{\partial \eta}(M\varphi) + \frac{1}{N}\frac{\partial^2}{\partial \eta^2}(V\varphi), \tag{3.3}$$

where

$$M = \frac{\eta(1-r)(1-\eta)-u}{\eta(r-1)-r}, \quad V = -\frac{1}{2}\frac{\eta[(1-\eta)(1+r)-u(1-2\eta)]}{\eta(r-1)-r}.$$

When $r = 1 + s/N$ with $s \ll N$, we have the following equation:

$$N\frac{\partial \varphi}{\partial T} = s\frac{\partial}{\partial \eta}(\eta(1-\eta)\varphi) + \frac{\partial^2}{\partial \eta^2}(\eta(1-\eta)\varphi).$$

This equation is studied in [Kimura (1994)]. In the case $s \ll 1$ the principal term in the expression for $\varphi(\eta, T)$ is proportional to $e^{-\mu_0 T}$, where

$$\mu_0 = \frac{1}{N}(1 + O((s)^2)).$$

This sets the typical time-scale of the process.

We can also study the case $1 \ll |s| \ll N$. In that limit, for $s > 0$, the region of interest is $\eta \ll 1$ (remember that $\eta = 0$ corresponds to all the "B" states). Thus the equation simplifies to

$$N\frac{\partial \varphi}{\partial T} = s\frac{\partial}{\partial \eta}(\eta\varphi) + \frac{\partial^2}{\partial \eta^2}(\eta\varphi).$$

This equation could be solved in terms of Laguerre polynomials, in general:

$$\varphi(\eta, T) = e^{-s\eta}\sum_{n=0}^{\infty} c_n L_n^1(s\eta) e^{-\frac{s(1+n)T}{N}}.$$

The Laguerre polynomials, $L_n^\alpha(x)$, satisfy the differential equation

$$\{x\frac{d^2}{dx^2} + (\alpha + 1 - x)\frac{d}{dx} + n\}L_n^\alpha(x) = 0.$$

Note that the leading transient gives $\mu_0 = \frac{|s|}{N} = |1 - r|$ in this limit. One could similarly treat the case of $s < 0, |s| \ll 1$. In general, $\mu_0 = \frac{1}{N} f(s)$ where $f(s) = 1 + O(s^2)$ for small s, but $f(s) \approx |s|$ for large s.

Absorption time. The method presented above provides a lot of information about the process. However, we can address some interesting questions without such a detailed description. For example, if we are only interested in the time it takes for a mutant of type "B" to appear and invade the population, we can do this directly, by looking at the absorption time for the Markov process. If we denote the number of elementary events until absorption starting from state i as T_i, we have

$$T_i = N + \sum_{m=0}^{N-1} P_{im} T_m, \quad 0 \leq i \leq N - 1. \tag{3.4}$$

The absorption time is then given by T_0. Solving system (3.4) directly is cumbersome, so we will use some approximations.

There are two processes that go on in the system: mutation and selection. If the characteristic time scales of the two processes are vastly different, our task of finding the absorption time simplifies greatly. Let us assume that u is very small, so that once a mutant of type "B" is produced, it typically has time to get fixated or die out before a new mutation occurs. In other words, once a mutant is produced, it is safe to assume that during its life-time no other mutations occur. In this case of rare mutations, the inverse time to absorption is roughly u times the probability to get absorbed in the state $b = N$ from the state $b = 1$ assuming $u = 0$.

For $u = 0$, the system has two absorbing states, $b = 0$ and $b = N$. Let us denote the probability to get absorbed in $b = N$ starting from the state $b = i$ as π_i. Then we have approximately

$$\frac{1}{t_{abs}} = N u \pi_1, \tag{3.5}$$

where the quantity π_1 is given by the system:

$$\pi_i = P_{iN} + \sum_{m=1}^{N-1} P_{im} \pi_m; \tag{3.6}$$

note that we set $u = 0$ in the expression for P. System (3.6) can be rewritten as

$$-\pi_{i-1} + (r + 1)\pi_i - r\pi_{i+1} = 0, \qquad 1 < i < N - 1,$$

where we canceled the common multiplier in terms of the matrix $I - P$ in the same row. The boundary conditions are

$$(r+1)\pi_1 - r\pi_2 = 0,$$
$$-\pi_{N-2} + (r+1)\pi_{N-1} = r.$$

We can look for a solution in the form $\pi_i = \alpha^i$. The quadratic equation for α gives the roots $\alpha = 1/r$ and $\alpha = 1$. Substituting $\pi_i = Ar^{-i} + B$ into the boundary conditions we obtain the solution,

$$\pi_i = \frac{r^{N-i}(1-r^i)}{1-r^N}. \tag{3.7}$$

Let us reserve the notation ρ for the quantity π_1:

$$\rho \equiv \pi_1 = \frac{1-1/r}{1-1/r^N}. \tag{3.8}$$

We have from (3.5):

$$\frac{1}{t_{abs}} = Nu\rho. \tag{3.9}$$

The same result is obtained if we solve system (3.4) explicitly and then take the first term in the Taylor expansion of T_0 in u.

In order for approximation (3.5) to be valid, we need to make sure that the time-scale related to mutation ($(Nu)^{-1}$) is much longer than the time-scale of the fixation/extinction processes. Only the fraction ρ of all mutants will successfully reach fixation, whereas the rest will be quickly driven to extinction. In order for each mutant lineage to be treated independently, we need to require that the time it takes to produce a *successful* mutant, $(\rho Nu)^{-1}$, is much larger than the typical time-scale of fixation, μ_0^{-1}. The value μ_0^{-1} is calculated above. We have the general expression,

$$u \ll \frac{\mu_0}{\rho N}.$$

In the case of neutral mutations, $\rho = 1/N$, $\mu_0 = 1/N$ and we arrive at the intuitive condition,

$$u \ll \frac{1}{N}, \qquad \text{if } |1-r| \ll \frac{1}{N}. \tag{3.10}$$

In the case where the mutation is positively or negatively selected, we have

$$u \ll (r^{N-1}N)^{-1}, \quad \text{if } r < 1, \quad \frac{1}{N} \ll |1-r| \ll 1, \tag{3.11}$$

$$u \ll r/N, \quad \text{if } r > 1, \quad \frac{1}{N} \ll |1-r| \ll 1. \tag{3.12}$$

The approximation of "almost absorbing" states. We will call a state of the system *homogeneous*, or pure, if all the N cells are of the same type. In the two-species model, these are the states $b = 0$ and $b = N$. States containing more than one type of cells $(1 < b < N)$ will be referred to as *heterogeneous*, or mixed states.

Since the mutation rate is very low relative to the absorption processes in the system (conditions (3.10-3.12)), the probability of finding the system in a heterogeneous state is very low. More precisely, the probability of finding the state with b cells of species "B" is of the order u for $1 < b < N$. The system spends most of the time in the states $b = 0$ and $b = N$. This allows us to make a further approximation of "almost absorbing" states.

Let us use the capital letters A and B for the probability to find the system in the state $b = 0$ and $b = N$ respectively. Strictly speaking, the state $b = 0$ is not absorbing, but it is long-lived. We have approximately, $A + B = 1$. Let us define the following "coarse-grained", continuous time stochastic process: the system jumps between two states, $A = 0$ and $A = 1$, with the following probabilities:

$$P(A=0, t+\Delta t | A=1, t) = u\rho\Delta t, \quad P(A=0, t+\Delta t | A=0, t) = 1,$$
$$P(A=1, t+\Delta t | A=1, t) = 1 - u\rho\Delta t, \quad P(A=1, t+\Delta t | A=0, t) = 0.$$

The Kolmogorov forward equations for this simple system can be written down, which describe the dynamics of the two-species model, (3.1):

$$\dot{A} = -uN\rho A \qquad A(0) = 1, \tag{3.13}$$

$$\dot{B} = uN\rho A, \qquad B(0) = 0, \tag{3.14}$$

where A is the probability to find the entire system in state "A", B is the probability to find the entire system in state "B" and ρ is given by equation (3.7). Equations (3.13-3.14) lead to the solution $A(t) = \exp(-uN\rho t)$ and $B = 1 - \exp(-uN\rho t)$.

A short-hand notation for coarse-grained differential equations (3.13-3.14) is as follows:

$$A \xrightarrow{uN\rho} B$$

We will use this notation later to describe the Kolmogorov forward equation of more complex mutation-selection networks.

3.2 A two-hit model

Now we will consider a two-hit model. In this section we will restrict ourselves to the Moran process (a constant population process). Later on, we will also discuss models with a changing population size.

3.2.1 *Process description*

We suppose that there are three types of cells: type "A", type "B" and type "C", and the mutation-selection network that governs the dynamics is as follows:

$$A_{(1)} \xrightarrow{\quad u \quad} B_{(r)} \xrightarrow{\quad u_1 \quad} C_{(r_1)} \tag{3.15}$$

The reproductive rates are respectively 1, r and r_1. As before, the reproductive rates must be interpreted as *relative* probabilities to be chosen for reproduction, rather than parameters defining the time-scale. We assume that type "A" can mutate into type "B" with probability u, and type "B" can mutate to type "C" with probability u_1. There are no other mutation processes in the system.

This model describes several biologically relevant situations. For instance, it may be directly applied for the two-hit hypothesis, that is, the process of the inactivation of a tumor suppressor gene. In the simplest case, the inactivation of the first allele of a tumor suppressor gene (TSP) does not lead to a phenotypic change, which corresponds to the value $r = 1$. This rigid definition can be relaxed to allow for certain *gene dosage effects.* For instance, the loss of one copy can lead to a certain change in the phenotype, and the loss of both copies will increase this effect. In this case, we could have $1 < r < r_1$. Finally, the case $r < 1$, $r_1 > 1$ means that the intermediate cell has a disadvantage compared to wild type cells. For example, this may correspond to the situation where the inactivation of the first allele is achieved by a large scale genomic alteration, such as a loss-of-heterozygocity event where many genes have been lost. This would lead to the intermediate product having a disadvantage compared to the wild type cells. Losing the remaining allele of the tumor suppressor gene will give the cell a growth advantage which may override the fitness loss of the previous event, resulting in $r_1 > 1$.

In general, the two-hit model described above refers to any two consecutive mutations, such that the first one may be positively or negatively selected (or neutral), and the second one confers a selective advantage to the cell.

Let us specify the states of the system by the variables a, b and c, which correspond to the number of cells of species "A","B" and "C", respectively. They satisfy the constraint $a + b + c = N$. We can characterize a state as a vector (b, c). In this notation, the state we start with is $(0, 0)$, which is all "A". The final state, which is the state of interest, is $(0, N)$, or all "C". The question we will study is again, the time of absorption in the state $c = N$.

We are interested in the case where the type "C" has a large selective advantage, i.e. $r_1 \gg (1, r)$, so that once there is one cell of type "C", this type will invade instantaneously with probability one. Under this assumption we can use a trick which allows us to view the dynamics as a one-dimensional process. Namely, let us consider the following reduced Markov process with the independent stochastic variable b: the states $b = i$ with $0 \leq i \leq N$ correspond to $a = N - i$, $b = i$, $c = 0$, and the state $b = N + 1$ contains all states with $c > 1$. The state $b = N + 1$ is absorbing, because we assume that once a mutant of type "C" appears, then cells "C" invade, so the system cannot go back to a state with $c = 0$. The transition probabilities are given by

$$P_{ij} = \begin{cases} \frac{u(N-i)+(1-u_1)ri}{\mathcal{N}_i} \frac{N-i}{N}, & j = i + 1, \\ \frac{(1-u)(N-i)}{\mathcal{N}_i} \frac{i}{N}, & j = i - 1, \\ \frac{u_1 ri}{\mathcal{N}_i}, & j = N + 1, \\ 1 - P_{i,i+1} - P_{i,i-1} - P_{i,N+1}, & j = i, \\ 0, & \text{otherwise}, \end{cases} \tag{3.16}$$

for $0 \leq i \leq N$, $P_{N+1,N+1} = 1$ and $P_{N+1,j} = 0$ for all $j \neq N + 1$. In some special cases, the absorption time can be found from equation (3.4), however, a direct solution is not possible in the general case, and we will use some approximations.

3.2.2 *Two ways to acquire the second hit*

Let us start from the all "A" state. If we are in the regime of homogeneous states, conditions (3.10-3.12), we can consider the lineages of each mutant of type "B" separately. Once a mutant of type "B" is created, it can either go extinct, or get fixated. A mutant of type "C" can be created before or after type "B" reaches fixation. This gives rise to two possible scenarios [Komarova *et al.* (2003)].

We will reserve the name *genuine two-step process* for a sequence of steps where starting from $(0,0)$, after some time the system finds itself in the state $(N,0)$ and then gets absorbed in the state $(0,N)$. In other words, starting from the all "A" state, the system gets to the state where the entire population consists of cells of type "B" and finally reaches fixation in the all "C" state.

We will use the term *tunneling* for such processes where the system goes from $(0,0)$ to $(0,N)$ without ever visiting state $(N,0)$. This means that from the all "A" state the system gets absorbed in the all "C" state, skipping the intermediate fixation of type "B".

It turns out that the computation of the waiting time for a mutant of type "C" to appear will be different in the two regimes. We start from looking at the tunneling regime and then talk about a genuine two-step process.

3.2.3 *The regime of tunneling*

The hazard function. We would like to calculate the probability, $P(t)$, that by the time t, at least one cell of type "C" has been produced, starting from all cells in the state "A" at time $t = 0$. This can be calculated by using the so-called *hazard function*, $h(t)$, which is defined as the probability to create a mutant of type "C" in the next interval Δt, given that it has not been produced so far. We have

$$P(t) = 1 - \exp\left(-\int_0^t h(t')dt'\right).$$

Let $\varphi_{j,k}(t)$ be the probability that at time t, we have $b = j$ and $c = k$. It is convenient to introduce the probability generating function,

$$\Psi(y,z;t) = \sum_{j,k} \varphi_{j,k}(t) y^j z^k. \tag{3.17}$$

The hazard function can be expressed in terms of Ψ as follows:

$$h(t) = -\frac{\dot{\Psi}(1,0;t)}{\Psi(1,0;t)}. \tag{3.18}$$

A trivial calculation shows that the probability $P(t)$ can be expressed in terms of the function Ψ as follows:

$$P(t) = 1 - \Psi(1, 0; t).$$

The meaning of function $\Psi(1, 0; t)$ is the probability that by time t, no cells of type "C" have been created.

Doubly stochastic process in a constant population. Let us calculate the function $\Psi(y, z; t)$. The initial condition of this problem is that all cells are normal (type "A"). In what follows we will assume that the time of interest is sufficiently short so that *most* cells remain type "A". In the Moran process, this means that $a \approx N$, and $b, c \ll N$. This assumption simplifies the problem. Following Moolgavkar, we will consider a filtered (or doubly-stochastic) Poisson process, where cells of type "B" are produced by mutations at the rate Nu (if we measure the time in terms of *generations*). Each cell of type "B" can produce a lineage (a clone). These lineages are independent of each other, so that the numbers of offspring of each "initial" cell are independent identically distributed random variables. Note that the assumption of the independence of the lineages breaks down as soon as a mutant of type "B" gets fixated. This happens with the probability ρ, see equation (3.8).

Assuming that fixation does not take place (the tunneling regime), we can write down the probability distribution, $\zeta_{j,k}$, which is the probability to have j cells of type "B" and k cells of type "C" starting from one cell of type "B" and no further "A"→"B" mutations. The corresponding probability generating function is given by

$$\Phi(y, z; t) = \sum_{j,k} \zeta_{j,k}(t) y^j z^k.$$

According to Parzen [Parzen (1962)], we have

$$\Psi(y, z; t) = \exp\left[-uN \int_0^t (1 - \Phi(y, z; t'))\, dt'\right]. \qquad (3.19)$$

The function $R(t) \equiv 1 - \Phi(y, z; t)$, which we call the *tunneling rate*, can be calculated directly. We start by writing down the Kolmogorov forward equation for $\zeta_{j,k}$:

$$\dot{\zeta}_{j,k} = \zeta_{j-1,k}(j-1)r(1-u_1) + \zeta_{j+1,k}(j+1) + \zeta_{j,k-1} j r u_1 - \zeta_{j,k}(r+1).$$

Here we use matrix (3.16) with $u = 0$, measure time in terms of generations and assume that $a \approx N$. Rewriting this for $\Phi(y, z; t)$ we obtain

$$\dot{\Phi}(y, z; t) = \left(y^2 r(1 - u_1) + yzru_1 - (r + 1)y + 1\right) \frac{\partial \Phi(y, z; t)}{\partial y}.$$

We want to find $\Psi(1, 0; t)$. Setting $z = 0$, we can write the equation for characteristics, which is a Riccati equation,

$$\dot{y} = y^2 r(1 - u_1) - (r + 1)y + 1.$$

Following the standard method, we change the variables, $y = -\frac{1}{2(1-u_1)} \frac{\dot{z}}{z}$, and obtain a second order linear equation for $z(t)$. Using the initial condition $\Phi(y, 0; 0) = y$, we can write down the solution:

$$\Phi(1, 0; t) = -\frac{b_1 + Ab_2 e^{(b_2 - b_1)t}}{r(1 - u_1)[1 + Ae^{(b_2 - b_1)t}]}, \tag{3.20}$$

where $A = -\frac{b_1 + r(1-u_1)}{b_2 + r(1-u_1)}$, and $b_1 > b_2$ are the roots of the quadratic equation,

$$b^2 + (r + 1)b + r(1 - u_1) = 0. \tag{3.21}$$

Behavior of the tunneling rate. There are three important limits:

(i) Disadvantageous intermediate step, $r < 1$, $|1 - r| \gg \sqrt{u_1}$. Then $b_1 = -r + \frac{u_1 r}{1-r}$, $b_2 = -1 - \frac{u_1 r}{1-r}$.
(ii) Neutral intermediate step, $|1 - r| \ll \sqrt{u_1}$. Then $b_1 = -1 + \sqrt{u_1}$, $b_2 = -1 - \sqrt{u_1}$.
(iii) Advantageous intermediate step, $r > 1$, $|1 - r| \gg 2\sqrt{u_1}$. Then $b_1 = -1 + \frac{ru_1}{r-1}$, $b_2 = -r - \frac{ru_1}{r-1}$.

Let us plot the tunneling rate, $R(t) = 1 - \Phi(1, 0; t)$, equation (3.20), as a function of time, see Figure 3.1. It starts at zero at $t = 0$, grows monotonically and reaches a saturation. We can identify three distinct regimes in the behavior of the function $\Phi(1, 0; t)$. For very short times, where

$$\max\{\sqrt{u_1}, |1 - r|\}t \ll 1,$$

we have the **linear regime**, where the function $R(t)$ grows linearly with time; we have

$$R(t) = ru_1 t.$$

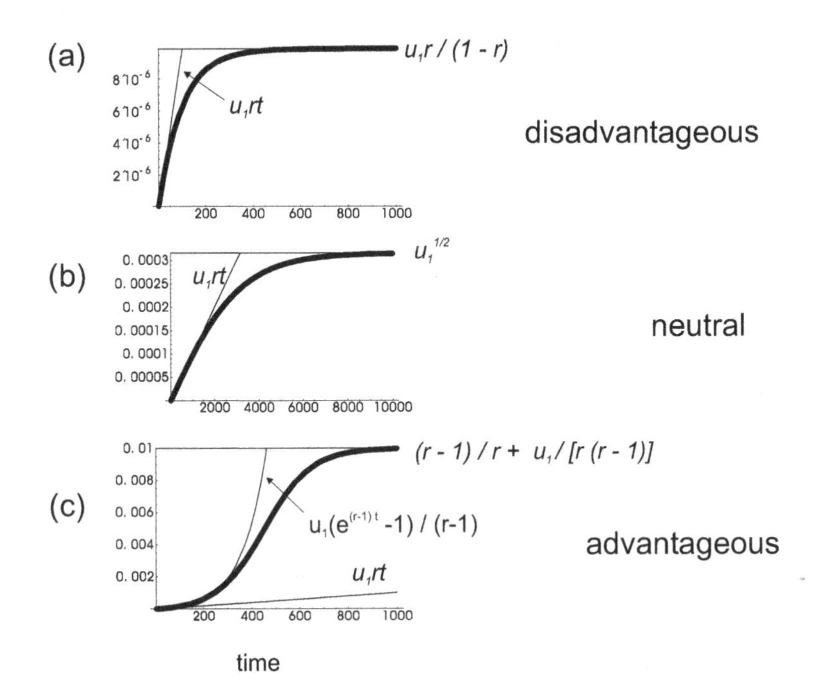

Fig. 3.1 The tunneling rate, $R(t) = 1 - \Phi(1, 0; t)$, equation (3.20), *vs* time, is represented by thick lines. There are three cases, (a) the intermediate mutant is disadvantageous, (b) it is neutral, and (c) it is advantageous. The linear regime and the regime of saturation are marked. Note that in (c), we have an intermediate, exponential growth regime. The parameter values are as follows: (a) $u = 10^{-7}$, $r = 0.99$, (b) $u = 10^{-7}$, $r = 0.999999$, (c) $u = 10^{-6}$, $r = 1.01$.

Next, during the times where $1/(r-1) < t < |\log u_1|/(r-1)$, we have the **intermediate regime** where the function $R(t)$ grows faster than linear. If we assume that the intermediate mutant is advantageous, such that $|r-1| \gg u_1$, then the expression for $R(t)$ can be simplified to give

$$R(t) = \frac{u_1 \left(e^{(r-1)t} - 1\right)}{r-1}.$$

As time increases, the function $\Phi(1, 0; t)$ quickly reaches **saturation**, that is, a steady-state, $\Phi(1, 0; t) = -b_1/[r(1-u_1)]$. In the case where the intermediate step is disadvantageous, $r < 1$, $|1-r| \gg \sqrt{u_1}$, the saturated value is

$$R(t) = \frac{u_1 r}{1-r},$$

where R is the "tunneling rate". In the case where the intermediate step is neutral, $|1-r| \ll \sqrt{u_1}$, the saturated value is

$$R = 1 - \Phi(1, 0; t) = \sqrt{u_1},$$

In the case where the intermediate step is advantageous, $r > 1$, $|1-r| \ll 2\sqrt{ru_1}$, the saturated value is

$$R = 1 - \Phi(1, 0; t) = \frac{r-1}{r} + \frac{u_1}{r(r-1)}.$$

Probability of double mutations for disadvantageous, neutral and advantageous intermediate mutants. Now we can use the detailed information about the behavior of the function $\Phi(1, 0; t)$ to evaluate the integral in (3.19) in different limiting cases. Let us first examine the case where the intermediate mutant is **disadvantageous**. Roughly speaking, the behavior of the function $1 - \Phi(1, 0; t)$ changes from linear to constant at $t = t_c$, where $t_c = 1/|1-r|$. At this value of time, the integral $\int_0^{t_c}(1 - \Phi(1, 0; t'))\,dt \sim 1$. In order to estimate the expression in (3.19), we need to know which of the regimes of growth makes the largest contribution. This is the same as to determine whether the exponent in (3.19) is negligible at the critical time, $t = t_c$. It is easy to see that if $uN \ll 1$, then there is a large contribution from the regime of saturation. On the other hand, for $uN \gg 1$, it is only the linear regime that contributes. Therefore we have the following answer:

$$P(t) = \begin{cases} 1 - exp\left(-\frac{Nuu_1r}{1-r}t\right), & uN \ll 1, \\ 1 - exp\left(-\frac{Nuu_1t^2}{2}\right), & uN \gg 1. \end{cases}$$

Next, we consider the case of a **neutral** intermediate mutant. The point of regime change is $t_c = 1/\sqrt{u_1}$. Again, if $uN \ll 1$, then there is a large contribution from the regime of saturation. On the other hand, for $uN \gg 1$, it is only the linear regime that contributes, and we have the result:

$$P(t) = \begin{cases} 1 - exp(-Nu\sqrt{u_1}t), & uN \ll 1, \\ 1 - exp\left(-\frac{Nuu_1t^2}{2}\right), & uN \gg 1. \end{cases}$$

Finally, we consider the case of **advantageous** mutants. There are three regimes. The two critical times where the regimes change are

$$t_1 = \frac{1}{r-1}, \quad t_2 = \frac{|\log u_1|}{r-1}.$$

The value of the function $\mathcal{R}(t) = \int_0^t (1 - \Phi(1,0;t'))\,dt$ can be estimated at these points by using the expression for the intermediate regime; we have

$$\mathcal{R}(t_1) = \frac{u_1(e-2)}{r-1} \sim u_1, \quad \mathcal{R}(t_1) = \frac{1 + u_1 \log u_1}{r-1} \sim 1.$$

Therefore, we have three regimes depending on the value of N: in the case where $N \ll 1/u$, the saturated regime contributes the most, for $1/u \ll N \ll 1/(uu_1)$, we have the contribution from the intermediate regime, and for $N \gg 1/(uu_1)$, we have the contribution from the linear regime only. In summary,

$$P(t) = \begin{cases} 1 - exp\left(-Nu\frac{r-1}{r}t\right), & N \ll 1/u, \\ 1 - exp\left(-Nuu_1 I(t)\right), & 1/u \ll N \ll 1/(uu_1), \\ 1 - exp\left(-\frac{Nuu_1 rt^2}{2}\right), & N \gg 1/(uu_1), \end{cases} \tag{3.22}$$

with $I(t) = \frac{1}{r-1}\left(\frac{e^{(r-1)t}-1}{r-1} - t\right)$ for the intermediate regime.

Applicability of the method. The method assumes independence of the lineages of the intermediate mutant. Thus for the method to work, the probability of fixation of intermediate mutants must be small compared to the probability of "tunneling". Therefore, the applicability is defined by the inequality,

$$\rho(r) < \lim_{t\to\infty} R,$$

where the right hand side is the saturated value of $1 - \Phi(1,0;t)$. This condition can be written as

$$N > N_{tun}, \tag{3.23}$$

where for disadvantageous intermediate mutants, we have:

$$N_{tun} = \frac{\log u_1 + 2\log\frac{r}{1-r}}{\log r}. \tag{3.24}$$

For neutral intermediate mutants, we have

$$N_{tun} = \frac{1}{\sqrt{u_1}}. \tag{3.25}$$

For advantageous mutants, we have

$$N_{tun} = \frac{\log \frac{(r-1)^2}{u1}}{\log r}. \tag{3.26}$$

Summary: tunneling rates. Having a high mutation rate, u_1 will increase the probability of tunneling. Also, in the case of a large population size, N, the fixation of type "B" becomes less probable thus making tunneling a more likely scenario. Finally, if type "B" is greatly disadvantageous, we also expect the system to tunnel from "A" to "B".

It is interesting that tunneling can be interpreted as making a two-hit process behaves effectively as a one-step process. Let us concentrate on the case where

$$N_{tun} < N < 1/u.$$

In the general case, we have the following diagram:

$$A \xrightarrow{R_{A\to C}} C.$$

This corresponds to the differential equations,

$$\dot{A} = -R_{A\to C}A \qquad A(0) = 1, \tag{3.27}$$
$$\dot{C} = R_{A\to C}A, \qquad C(0) = 0. \tag{3.28}$$

This is similar to the one-hit model, which we considered in the previous sections, see equations (3.13-3.14). The tunneling rate, $R_{A\to C}$, is different depending on whether the intermediate mutant, "B", is positively or negatively selected. We have three cases:

- **Type "B" negatively selected.** If $r < 1$ and $|1-r| \gg \sqrt{u_1}$, then we have tunneling from "A" to "C" with the rate

$$R_{A\to C} = \frac{Nuru_1}{1-r}. \tag{3.29}$$

- **Type "B" neutral.** If $|1-r| \ll \sqrt{u_1}$, then we have tunneling from "A" to "C" with the rate

$$R_{A\to C} = Nu\sqrt{u_1}. \tag{3.30}$$

- **Type "B" positively selected.** If $r > 1$ and $|1-r| \ll \sqrt{u_1}$, then we have tunneling from "A" to "C" with the rate

$$R_{A\to C} = Nu\frac{r-1}{r}. \tag{3.31}$$

3.2.4 *Genuine two-step processes*

If the number of cells in the population, N, is sufficiently small, then the assumption of the previous section (3.23) breaks down, and tunneling does not happen. In this case, the dynamics can be represented by the diagram

$$A \xrightarrow{\quad R_{A\to B} \quad} B \xrightarrow{\quad R_{B\to C} \quad} C$$

with

$$R_{A\to B} = Nu\rho, \qquad R_{B\to C} = Nu_1.$$

Here we assumed that $\frac{1-1/r_1}{1-1/r_1^N} \approx 1$. The corresponding differential equations are

$$\dot{A} = -Nu\rho A, \qquad A(0) = 1, \tag{3.32}$$

$$\dot{B} = Nu\rho A - Nu_1 B, \qquad B(0) = 0, \tag{3.33}$$

$$\dot{C} = Nu_1 B. \tag{3.34}$$

3.2.5 *Summary of the two-hit model with a constant population*

To put all the results together, we will describe the dynamics of the acquisition of a double mutant, as a function of time. Depending on the population size, the behavior is quite different. For small populations, where

$$N < N_{tun},$$

we have

$$P(t) = 1 - \frac{u_1 e^{-Nu\rho t} - u\rho e^{-Nu_1 t}}{u_1 - u\rho}.$$

This comes from solving equations (3.32-3.34) and setting $P(t) = C(t)$. The function ρ depends on the fitness of the intermediate mutant, $\rho = \frac{1-1/r}{1-1/r^N}$. The value N_{tun} is also defined by r, see formulas (3.24), (3.25) and (3.26).

For intermediate values on N, such that

$$N_{tun} < N < 1/u,$$

we have the following behavior:

$$P(t) = 1 - e^{-R_{A\to C}t},$$

which comes from the solution of equations (3.27-3.28). The rate of tunneling, $R_{A\to C}$ again depends on the fitness of the intermediate mutant, see formulas (3.29), (3.30) and (3.31).

Finally, for very large population sizes, such that

$$N > 1/u,$$

we have

$$P(t) = 1 - e^{-\frac{Nuu_1rt^2}{2}}.$$

For advantageous mutants we have another intermediate regime, which comes for $1/u < N < 1/(uu_1)$. This is given in equation (3.22).

3.3 Modeling non-constant populations

3.3.1 *Description of the model*

Consider the process described by the following mutation-selection network:

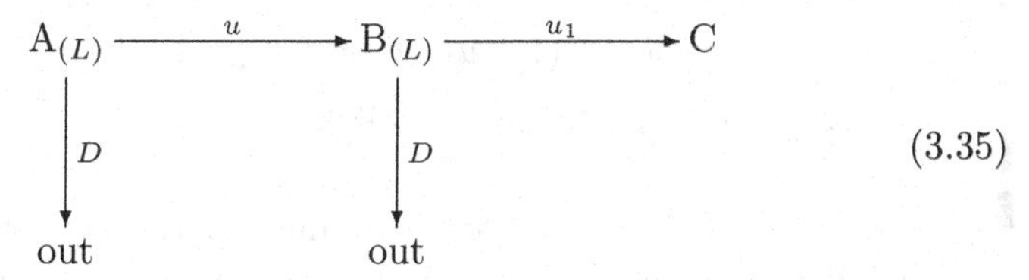

(3.35)

Here the reproductive rates of types "A" and "B" are the same and equal to L, and the death rates are D. Type "A" can mutate into type "B" with probability u, and type "B" can mutate to type "C" with probability u_1. There are no other mutation processes in the system. We assume that $u \ll u_1$, and that the dynamics follow a Poisson process, where in time interval Δt, the following events can occur:

- With probability $L(1-u)\Delta t$ a cell of type "A" reproduces, creating an identical copy of itself,
- With probability Lu a cell of type "A" reproduces with a mutation, creating a cell of type "B",
- With probability $L(1-u_1)\Delta t$ a cell of type "B" reproduces, creating an identical copy of itself,

- With probability Lu_1 a cell of type "B" reproduces with a mutation, creating a cell of type "C",
- With probability $D\Delta t$ a cell of type "A" dies,
- With probability $D\Delta t$ a cell of type "B" dies.

We start with one cell of type "A", and follow the process until the first cell of type "C" has been created. As before, we would like to calculate the probability, $P(t)$, that one cell of type "C" has been created as a function of time. Before, we used the approximation of a doubly-stochastic process, see equation (3.19). A similar approach can be used to describe expanding populations, as long as we can assume that the expansion process is nearly deterministic. On the other hand, if we want to take account of the stochasticity in the colony growth, we need to perform a more general calculation, which is described below. An example of a system where such a calculation is necessary is a colony which grows from very low numbers, such that at the beginning, stochastic effects define the growth (or death) of the cell population.

Let us consider the probability $\xi_{i,j,k}(t)$ that at time t, we have $a = i$, $b = j$ and $c = k$. We have the Kolmogorov forward equation,

$$\begin{aligned}\dot{\xi}_{i,j,k} &= \xi_{i-1,j,k}L(i-1)(1-u) + \xi_{i+1,j,k}D(i+1)\\ &+ \xi_{i,j-1,k}[L(j-1)(1-u_1) + Liu] + \xi_{i,j+1,k}D(j+1)\\ &- \xi_{i,j,k-1}Lju_1 - \xi_{i,j,k}(L+D)(i+j).\end{aligned} \tag{3.36}$$

Note that here we do not consider the dynamics of the double-mutants: once produced, they remain in the colony. Birth-death processes of double-mutants can be incorporated leading to a slightly more complicated system. Let us define the generating function

$$\Psi(x,y,z;t) = \sum_{i,j,k} \xi_{j,k}(t)x^i y^j z^k. \tag{3.37}$$

The quantity $\Psi(1,1,0;t)$ has the meaning of the probability that at time t, no cells of type "C" have been created. The quantity in question, the probability that at least one cell of type "C" has been created by time t, is given by

$$P_2(t) = 1 - \Psi(1,1,0;t).$$

The subscript "2" refers to the number of hits (from "A" to "B" and from "B" to "C"). The function $\Psi(x, y, z; t)$ satisfies the following equation:

$$\frac{\partial \Psi}{\partial t} = \frac{\partial \Psi}{\partial x}[x^2 L(1-u) + D + yxLu - (L+D)x] + \frac{\partial \Psi}{\partial y}[y^2 L(1-u_1) + D + zyLu_1 - (L+D)y]. \quad (3.38)$$

The equations for characteristics are:

$$\dot{x} = L(1-u)x^2 + [Luy - (L+D)]x + D, \quad (3.39)$$
$$\dot{y} = L(1-u_1)y^2 + [Lu_1 z - (L+D)]y + D, \quad (3.40)$$
$$\dot{z} = 0 \quad (3.41)$$

(note that the last equation is trivial because we suppress the dynamics of double-mutants; if we include their dynamics, the equation for z would reflect that). We want to obtain the expression for $\Psi(1, 1, 0; t)$, thus we can set the initial conditions

$$x(0) = 1, \qquad y(0) = 1, \qquad z(0) = 0.$$

We obtain immediately from equation (3.41) that $z = 0$.

3.3.2 *A one-hit process*

First, we consider a simplified model with only one hit. Note that the function

$$P_1(t) = 1 - y(t)$$

with the initial condition as above, has the meaning of the probability that a cell of type "C" has been created starting with one cell of type "B". The corresponding diagram is this:

$$\begin{array}{ccc} \mathrm{B}_{(L)} & \xrightarrow{u_1} & \mathrm{C} \\ \downarrow D & & \\ \text{out} & & \end{array} \quad (3.42)$$

The equation for $y(t)$, (3.40), can then be solved exactly with $z = 0$. We set

$$y = -\frac{\dot{Y}}{L(1-u_1)Y},$$

and obtain a Riccati equation for Y:

$$\ddot{Y} + (L+D)\dot{Y} + L(1-u_1)DY = 0.$$

The solution can be easily obtained:

$$y = -\frac{b+1+Ab_2e^{(b_2-b_1)t}}{L(1-u_1)(1+Ae^{(b_2-b_1)t})},$$

where

$$A = -\frac{b_1 + L(1-u_1)}{b_2 + L(1-u_1)},$$

and $b_1 > b_2$ are roots of the quadratic equation,

$$b^2 + (L+D)b + L(1-u_1)D = 0.$$

This is similar to equation (3.21). The difference is that now, we allow for expansion and contraction of the population by using different values for L and D.

3.3.3 *Three types of dynamics*

Again, there are three limits in this problem.

Slow expansion or contraction. In the limit where $|L-D| \ll 2\sqrt{LDu_1}$, we have

$$b_1 = -L(1-\sqrt{u_1}), \qquad b_2 = -L(1+\sqrt{u_1}),$$

and the behavior of y is as follows:

- for small t, that is when $L\sqrt{u_1}t \ll 1$, we have $y = 1 - Lu_1t$, whereas
- for larger t such that $L\sqrt{u_1}t > 1$, $y(t) \to y_\infty$, where

$$y_\infty = 1 - \sqrt{u_1}.$$

Shrinking population. If $D > L$ and $D - L \gg 2\sqrt{LDu_1}$, we have

$$b_1 = -L - \frac{LDu_1}{L-d}, \quad b_2 = -D + \frac{LDu_1}{L-D},$$

and the solution has the following shape:

- $y(t) = 1 - u_1 Lt$ for $(D-L)t \ll 1$, as before, and
- for t such that $(D-L)t > 1$, $y(t)$ tends to a constant,

$$y_\infty = 1 - \frac{Lu_1}{D-L}.$$

Fast expansion. In the opposite case where $L - D \gg 2\sqrt{LDu_1}$, we have

$$b_1 = -D + \frac{LDu_1}{L-D}, \quad b_2 = -L - \frac{LDu_1}{L-d},$$

and the solution has the following shape:

- $y(t) = 1 - u_1 Lt$ for $(L-D)t \ll 1$, as before, and
- for t such that $(L-D)t > 1$, $y(t)$ saturates at a different value,

$$y_\infty = \frac{D}{L}\left(1 - \frac{Du_1}{L-D}\right).$$

3.3.4 *Probability to create a mutant of type "C"*

- For slow expansion/contraction, initially ($L\sqrt{u_1}T \ll 1$) we have a linear growth, $P_1(t) \approx Lu_1 t$, and then the probability saturates at $P_1 = \sqrt{u_1}$.
- For a shrinking population, initially ($(D-L)t \ll 1$) we have $P_1(t) \approx Lu_1 t$, and then it saturates at $P_1 = Lu_1/(D-L)$.
- For an expanding population, we have initially (when $(L-D)t \ll 1$) $P_1(t) \approx Lu_1 t$, and then $P_1 = 1 - D/L$.

The function $P_1(t)$ is presented in Figure 3.2, the upper line.

3.3.5 *A two-hit process*

Next, we can solve the equation for x, equation (3.39), in order to obtain

$$P_2(t) = 1 - x(t),$$

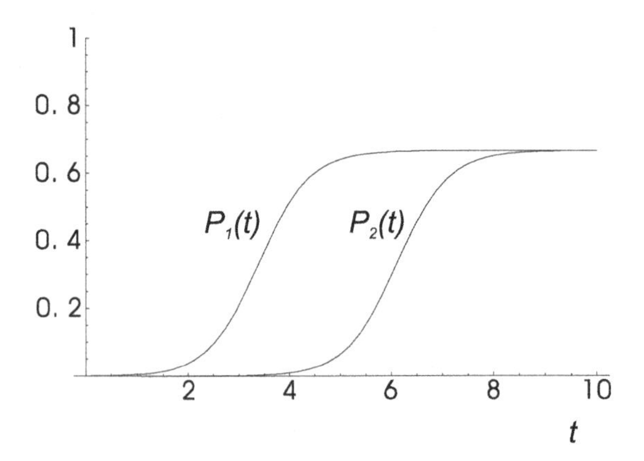

Fig. 3.2 The functions $P_1(t)$ and $P_2(t)$ for an expanding population, corresponding to one-hit and two-hit processes. The function $P_i(t)$ denotes the probability to create at least one "dangerous" mutant by time t, in an i-hit process. The parameter values are $L = 3$, $D = 1$, $u = u_1 = 5 \times 10^{-4}$.

the probability that a mutant of type "C" has been created starting from one cell of type "A". Using the same change of variables,

$$x = -\frac{\dot{Y}}{L(1-u)X},$$

we again obtain a Riccati equation,

$$\ddot{X} + (L + D - Luy)\dot{X} + L(1-u)DX = 0, \tag{3.43}$$

but now it contains a time-dependent coefficient. If we substitute the value of $y(t)$ at $t = 0$, $y = 1$, we will get the solution

$$X(t) = e^{-L(1-u)t}, \qquad x(t) = 1.$$

For the saturated value of $y(t)$, $y(t) = D/L$, we obtain

$$x(t) = 1 - (L - D)ut, \quad (L - D)t \ll 1, \tag{3.44}$$

$$x(t) = D/L, \quad (L - D)t > 1. \tag{3.45}$$

This gives us bounds for $P_2 = 1 - x(t)$. In general, we can solve equation (3.43) numerically. For a particular set of parameters, the function $P_2(t)$ is presented in Figure 3.2.

3.4 Overview

In this chapter we developed the main ideas of the stochastic formalism for two-hit models of carcinogenesis. In the next chapters we will see how these ideas can be applied to studying some of the most intriguing questions of cancer initiation and progression.

Chapter 4

Microsatellite and chromosomal instability in sporadic and familial cancers

This is the first of a number of chapters which investigates the relationship between carcinogenesis and genetic instability. Here, we will examine the most basic scenario: the generation of the first malignant cell. Does the presence of genetic instability result in a faster generation of the first malignant cell? The mathematics in this chapter are applications of the formalisms developed in Chapter 3. Here we present a simple example of how stochastic models developed for two-hit processes can be applied to biological reality.

We will concentrate on cancers which are initiated via the inactivation of a tumor suppressor gene. That is, both the maternal and the paternal copy of the gene have to lose function. A particular example which will be discussed in this context is colorectal cancer. Colorectal cancer is a major cause of mortality in the Western world. Approximately 5% of the population develop the disease, and about 40% of those diagnosed with it die within 5 years. Considerable progress has been made in identifying genetic events leading to colorectal cancer. Somatic inactivation of the adenomatous polyposis coli (APC) gene is believed to be one of the earliest steps occurring in sporadic colorectal cancer. It has been observed that the frequency of APC mutations is as high in small lesions as it is in cancers. Evidence that the APC gene plays a crucial role in colorectal cancer also comes from the study of individuals with familial adenomatous polyposis coli (FAP). FAP patients inherit a mutation in one of the copies of the APC gene; by their teens, they harbor hundreds to thousands of adenomatous polyps.

The APC gene is a tumor suppressor gene which controls cell birth and cell death processes. Inactivation of only one copy of the APC gene does not seem to lead to any phenotypic changes. Inactivation of both copies of this gene appears to result in an increased cell birth to death ratio in the corresponding cell and leads to clonal expansion and the formation of a *dysplastic crypt.* Here, we define a dysplastic crypt as a crypt that consists of cells with both copies of the APC gene inactivated. Dysplastic crypts are at risk of developing further somatic mutations which will eventually lead to cancer. The typical estimate is that an average 70 year old has about 1-10 dysplastic crypts, but precise counts have never been published.

How can the tumor suppressor gene be inactivated? A point mutation can induce a loss of function in one copy of the gene. Both copies of the gene can be inactivated by two subsequent point mutations in the same cell: one in the maternal, and the other in the paternal allele. Each mutation would occur with the physiological mutation rate of 10^{-7} per gene per cell division. Now consider genetic instability. As explained in Chapter 1, there are two major types of instabilities [Lengauer *et al.* (1998); Sen (2000)]: (i) small scale subtle sequence changes, such as microsatellite instability (MSI). The MSI phenotype is generated if specific MSI genes are inactivated. Both copies of an MSI gene need to be mutated. MSI basically results in an elevated point mutation rate in the context of repeat sequences called microsatellites. (ii) Gross chromosomal alterations can occur, and this is known as chromosomal instability (CIN). The genetic basis of CIN is uncertain, and specific scenarios will be discussed below. If one copy of the tumor suppressor gene has been inactivated by a point mutation, the other copy can be inactivated very quickly in CIN cells due to the loss of the healthy allele. This can occur through a variety of mechanisms. They include loss of the remaining chromosome and loss of part of the remaining chromosome. These processes are also called *loss of heterozygocity*, or *LOH.*

While genetic instability might speed up the loss of tumor suppressor function, the MSI or CIN phenotypes need to be generated first (for example by basic point mutations). This chapter discusses a mathematical analysis of how MSI and CIN influence the rate of tumor suppressor gene inactivation. We will apply this analysis to various scenarios which include the sporadic (spontaneous) development of colon cancer, and familial colon cancers. We start with some more detailed biological facts about CIN and MSI in colon cancer and then present the mathematical analysis.

4.1 Some biological facts about genetic instability in colon cancer

Here we will study the role that CIN and MSI may play in the inactivation of the APC gene. About 13 % of all colorectal cancers have MSI and most of the rest are characterized by CIN [Lengauer *et al.* (1998)]. MSI occurs in virtually all hereditary non-polyposis colorectal cancers (HNPCC), which account for about 3% of all colorectal cancers. The MSI phenotype results from defective mismatch repair. Several genes have been identified whose inactivation leads to an increased rate of subtle genetic alterations. The main ones are hMSH2 and hMLH1. Both copies of an MSI gene must be inactivated in order for any phenotypic changes to occur. HNPCC patients inherit a mutation in one of the copies of an MSI gene and normally develop colorectal tumors in their forties. Unlike FAP patients, they do not have a vastly increased number of polyps, but the rate of progression from polyp to cancer is faster.

Molecular mechanisms leading to CIN in human cancers remain to be understood. It has been proposed that CIN might be caused by mutations in genes involved in centrosome/microtubule dynamics, or checkpoint genes that monitor the progression of the cell cycle, e.g. the spindle checkpoint or the DNA-damage checkpoint [Kolodner *et al.* (2002)]. For example, heterozygous mutations in the mitotic spindle checkpoint gene hBUB1 have been detected in a small fraction of colorectal cancers with the CIN phenotype [Cahill *et al.* (1998); Gemma *et al.* (2000); Imai *et al.* (1999); Ohshima *et al.* (2000)]. Also, the MAD2 gene seems to be transcriptionally repressed in various solid tumors [Li and Benezra (1996); Michel *et al.* (2001); Ro and Rannala (2001); Wang *et al.* (2000)]. Some CIN genes might act in a dominant-negative fashion: an alteration in one allele leads to CIN.

4.2 A model for the initiation of sporadic colorectal cancers

The colonic epithelium is organized in crypts covered with a self-renewing layer of cells (Figure 4.1). The total number of crypts is of the order of $M = 10^7$ in a human. Each crypt contains of the order of 10^3 cells. A crypt is renewed by a small number of stem cells (perhaps $1 - 10$) [Ro and Rannala (2001); Yatabe *et al.* (2001)]. The life cycle of stem cells is of the order of $1 - 20$ days [Bach *et al.* (2000); Potten *et al.* (1992)]. Stem cells

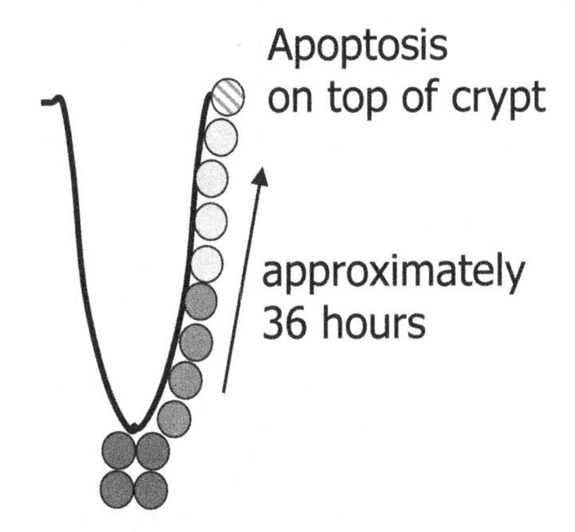

Fig. 4.1 The epithelium of the colon is organized into crypts. Each crypt contains about 10^3 cells. A small number of (stem) cells, which are thought to be located at the bottom of the crypt, divide asymmetrically to replenish the whole crypt. They give rise to differentiated cells which travel within 36 hours to the top of the crypt where they undergo apoptosis. Inactivation of both copies of the APC gene is believed to prevent apoptosis. The mutated cells remain on the top of the crypt, continue to divide and ultimately take over the crypt. This process gives rise to a dysplastic crypt, which represents the first step on the way to colorectal cancer.

give rise to differentiated cells which divide at a faster rate, and travel to the top of the crypt where they undergo apoptosis.

We start with the basic model of sporadic colorectal cancer initiation [Komarova *et al.* (2002)]. All the relevant parameters with their respective values are summarized in Table 4.1. Let us assume that the *effective population size* of a crypt is N; this means that N cells are at risk of developing mutations which can lead to cancer. The value of N is unknown. As will be explained in detail in Chapter 5, one hypothesis is that only the stem cells are at risk of developing cancer, which gives $N \sim 1 - 10$, and in this case the average turnover rate would be $\tau = 1 - 20$ days. Alternatively, we could assume that some differentiated cells are also at risk. In this case, N might be of the order of 100 and the average turnover rate could be less than 1 day. Here we will concentrate on the model with $N \sim 1 - 10$; some implications of the other model will also be discussed. In this chapter, we will not consider the details of the population structure. That is, the distinction between stem cell and differentiated cell division patterns will be ignored. Chapter 5 suggests a way of incorporating this in the model.

Table 4.1 Parameters, notations and possible numerical values; the mutation and LOH rates are given per gene per cell division.

Quantity	*Definition*	*Range*
M	Number of crypts in a colon	10^7
N	Effective number of cells in a crypt	$1 - 100$
τ	Effective time of cell cycle, days	$1 - 20$
u	Probability of mutation in normal (non-MSI) cells	10^{-7}
$\tilde{u}$	Probability of mutation in MSI cells	10^{-4}
p_0	Rate of LOH in normal (non-CIN) cells	10^{-7}
p	Rate of LOH in CIN cells	10^{-2}
n_m	Total number of MSI genes	$2 - 5$
n_c	Total number of CIN genes	?

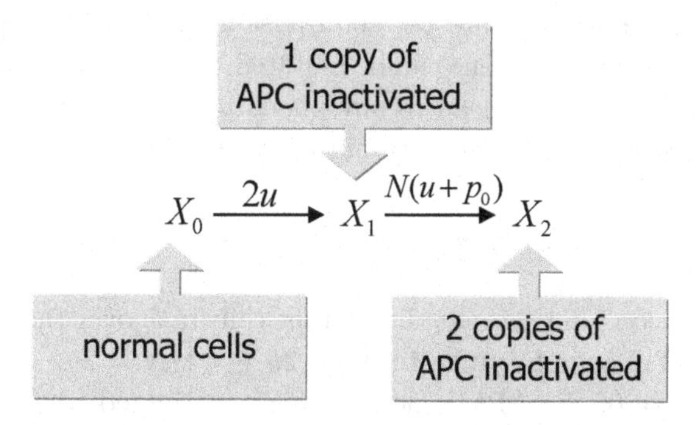

Fig. 4.2 Mutation-selection network of sporadic colorectal cancer initiation. Initially, the crypt is at the state X_0, i.e. all cells are wild-type. With the rate $2u$, cells with one copy of the APC gene mutated will take over the crypt (state X_1). This rate of change is calculated as N times the probability (per cell division) to produce a mutant of X_1 ($2u$ because either of the two alleles can be mutated) times the probability of *one* mutant of type X_1 to get fixed ($1/N$ since there is no phenotypic change). From state X_1 the system can go to state X_2 (both copies of the APC gene inactivated) with the rate $N(u + p_0)$. This rate is calculated as N times the probability per cell division to produce a mutant of X_2 (u for an independent point mutation plus p_0 for an LOH event) times the probability of the *advantageous* mutant of type X_2 to take over (this is 1).

Let us denote by X_0, X_1 and X_2 the probability that the whole crypt consists of cells with 0, 1 and 2 copies of the APC gene inactivated, respectively. The simplest mutation-selection network leading from X_0 to X_1 to X_2 is shown in Figure 4.2. The rate of change is equal to the probability that one relevant mutation occurs times the probability that the mutant cell will take over the crypt.

In the beginning (see Figure 4.2), all cells are wild type. The fist copy of the APC gene can get inactivated by a mutation event. Because the mutation rate per gene per cell division, $u \approx 10^{-7}$, is very small and the number of cells, N, is not large, it is safe to assume that once a mutation occurs, the population typically has enough time to become homogeneous again before the next mutation occurs. The condition is that the mutation rate, u, is much smaller than $1/N$, as was derived in Chapter 3. This means that most of the time, the effective population of cells in a crypt can be considered as homogeneous with respect to APC mutations. Under this assumption we have $X_0 + X_1 + X_2 = 1$.

Initially, all the N cells of a crypt have two copies of the APC gene. The first copy of the APC gene can be inactivated by means of a point mutation. The probability of mutation is given by N (a mutation can occur in any of the N cells) times the mutation rate per cell division, u, times 2, because any of the two copies of the APC gene can be mutated. Because inactivation of one copy of the APC does not lead to any phenotypic changes, the rate of fixation of the corresponding (neutral) mutant is equal to

$$\rho = \lim_{r \to 1} \frac{1 - 1/r}{1 - 1/r^N} = 1/N,$$

see Chapter 3. "Fixation" means that the mutant cells take over the crypt. Therefore, the rate of change from X_0 to X_1 is $2uN \times 1/N = 2u$.

Once the first allele of the APC gene has been inactivated, the second allele can be inactivated either by another point mutation or by an LOH event. This process occurs with rate $N(u + p_0)$, where p_0 is the rate of LOH in normal (non-CIN) cells. We assume that mutants with both copies of the APC gene inactivated have a large selective advantage, so that once such a mutant is produced, the probability of its fixation is close to one. This assumption is made for simplicity. More generally, the relative fitness of type X_2 is $\tilde{r}$, whereas the fitness of type X_0 and X_1 is 1. Then the second rate in Figure 4.1 should be taken to be $N\tilde{\rho}(u + p_0)$, with $\tilde{\rho} = (1 - 1/\tilde{r})/(1 - 1/\tilde{r}^N)$. If the population size is not too large, and the relative fitness of type X_2 is much greater than 1, we have $\rho_2 \to 1$, and we obtain the expression $N(u + p_0)$.

The mutation-selection network of Figure 4.2 is equivalent to a linear system of ordinary differential equations (ODE's), where the rates by the arrows refer to the coefficients and the direction of the arrows to the sign of the terms. One (non-dimensional) time unit ($t/\tau = 1$) corresponds to a generation turn-over. The calculations leading to the mutation-selection network are performed for a Moran process where the population size is kept constant by removing one cell each time a cell reproduces, see Chapter 3. Our biological time-unit again corresponds to N "elementary events" of the Moran process, where an elementary event includes one birth and one death. We have:

$$\dot{X}_0 = -2uX_0,$$
$$\dot{X}_1 = 2uX_0 - N(u + p_0)X_1,$$

with the constraint $X_0 + X_1 + X_2 = 1$ and the initial condition

$$X_0(0) = 1, \quad X_1(0) = 0.$$

Here, we use the fact that the intermediate mutant is neutral and that the population size is small ($N < N_{tun}$, see Chapter 3) so that stochastic tunneling does not take place. Calculations for larger values of N can also be performed.

Using $ut/\tau \ll 1$ and $N(p_0+u)t/\tau \ll 1$, we can approximate the solution for X_2 as

$$X_2(t) = Nu(u + p_0)(t/\tau)^2.$$

The quantity $X_2(t)$ stands for the probability that a crypt is dysplastic (i.e. consists of cells with both copies of the APC gene inactivated) at time t measured in days. This formula is a consequence of the fact that in the parameter regime we are considering, $APC^{-/-}$ cells are produced as a result of a genuine two-hit process (see Chapter 3). There are two steps that separate the state X_0 from the state X_2, and thus the expected number of dysplastic crypts in a person of age t is proportional to the product of the two rates and the second power of time. This reminds us of the general Armitage-Doll model where the power dependence of the probability of cancer is equal to the number of mutations in the multi-stage process. In our case, the number of mutations needed to create a dysplastic crypt is two.

The probability to have i dysplastic crypts by the age t is given by a simple binomial, $\binom{M}{i}X_2(t)^i(1 - X_2(t))^{M-i}$. The expected number of dysplastic crypts in a person of age t is then given by the following quantity,

$$MNu(u + p_0)(t/\tau)^2. \tag{4.1}$$

Some estimates of the expected number of dysplastic crypts, based on equation (4.1), are given in Table 4.5.

Table 4.2 **Sporadic colorectal cancer**: the expected number of dysplastic crypts, at 70 years of age, the simple model. $M = 10^7$, $N = 5$, $u = 10^{-7}$ and $t = 70$ years.

	$p_0 = 10^{-7}$	$p_0 = 10^{-6}$
$\tau = 1$	654	3,595
$\tau = 3$	73	400
$\tau = 10$	7	36
$\tau = 20$	2	9

One has to be careful when comparing these calculations with data. It is possible that dysplastic crypts can be lost. The model presented here gives the number of dysplastic crypts that are being *produced* over time, which could be larger than the actual number of dysplastic crypts that patients have at a particular time point. Exact measurements of the incidence of dysplastic crypts will provide important information about the crucial parameters of colorectal cancer initiation.

4.3 Sporadic colorectal cancers, CIN and MSI

Let us now consider the possibility of developing genetic instabilities during cancer initiation. Starting from a population of normal cells, three different events can occur: (i) inactivation of the first copy of the APC gene, (ii) inactivation of the first copy of one of n_m MSI genes, and (iii) mutation of one copy of one of n_c CIN genes.

We use the notation X_i, Y_i and Z_i, respectively, for the probability that a crypt consists of normal cells, CIN cells or MSI cells with i copies of the APC gene inactivated, see Table 4.3. Figure 4.3 shows the mutation-selection network of colorectal cancer initiation including CIN and MSI. All the transition rates are calculated as the relevant mutation rate times the probability that the mutant will take over the crypt.

Let us denote the rate of LOH in CIN cells as p. We assume that the crucial effect of CIN is to increase the rate of LOH [Bardelli *et al.* (2001); Lengauer *et al.* (1997)], which implies $p > p_0$. Intuitively, the advantage of CIN for the cancer cell is to accelerate the loss of the second copy of a tumor suppressor gene. Similarly, the advantage of MSI is to increase the

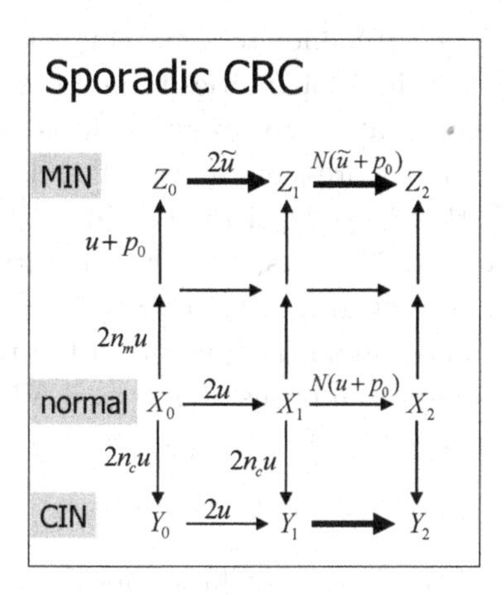

Fig. 4.3 Mutation-selection network of sporadic colorectal cancer initiation including CIN and MSI. From the initial wild-type state, X_0, the crypt can change to state X_1 as in Figure 4.1, acquire a CIN mutation (the arrow down) or an MSI mutation (the arrow up). The line $X_0 \to X_1 \to X_2$ is identical to the process in Figure 4.1 of developing a dysplastic crypt with no genetic instabilities. The bottom row of the diagram corresponds to CIN cells acquiring the first, and then the second, mutation (loss) of the APC gene; the second copy can be lost by a point mutation or by an LOH event whose rate is much larger for CIN cells than it is for normal or MSI cells. The state Y_2 corresponds to a CIN dysplastic crypt. The top row is the development of an MSI dysplastic crypt. The MSI phenotype is characterized by an increased point mutation rate, $\tilde{u}$. The state Z_2 is an MSI dysplastic crypt. Red arrows denote faster steps. Note that it takes only one leap (down) to go to a CIN state from a state with no genetic instability, because CIN genes are dominant-negative. It takes two steps to acquire MSI (up) because both copies of an MSI gene need to be inactivated before any phenotypic changes happen.

Table 4.3 The three major classes of homogeneous states.

Quantity	*Definition*	*Point mutation rate*	*Rate of LOH*
X_0, X_1, X_2	non-CIN, non-MSI	u	p_0
Y_0, Y_1, Y_2	CIN	u	$p > p_0$
Z_0, Z_1, Z_2	MSI	$\tilde{u} > u$	p_0

point-mutation rate, which means that $\tilde{u} > u$.

We are interested in the probability to find the crypt in the state X_2, Y_2 and Z_2 as a function of t. In other words, we want to know the probability for the dysplastic crypt to have CIN (Y_2), MSI (Z_2) or no genetic instability (X_2). The mutation-selection network of Figure 4.3 is more complicated than the one-dimensional network of Figure 4.2, but the solutions for X_2, Y_2 and Z_2 can still be written down.

The diagram of Figure 4.3 corresponds to a system of 11 linear ODE's describing the time-evolution of the probabilities to find the system in any of the 12 possible homogeneous states. An exact solution can be written down but it is a very cumbersome expression, so we will make some approximations. Let us use the fact that the quantities ut/τ and $N(u+p_0)t/\tau$ are very small compared to 1 for $t \sim 70$ years, and the quantity $Npt/\tau \gg 1$. This tells us that the steps in the diagram characterized by the rates u and p_0 are slow (rate limiting) compared to the steps with the rate p. Taking the Taylor expansion of the solution in terms of ut/τ and $N(u+p_0)t/\tau$, we obtain the following result:

$$X_2(t) = Nu(u+p_0)(t/\tau)^2, \qquad Y_2(t) = 4n_c u^2 (t/\tau)^2. \tag{4.2}$$

The rate $\tilde{u}$ is neither fast nor slow, so the solution for Z_2 is more complicated. We have

$$Z_2(t) = \frac{n_m u(u+p_0)}{(ab\tilde{u})^2(a-b)} \Big(2(b^3 E_a - a^3 E_b + a^3 - b^3 + ab(b^2-a^2)\tilde{u}t/\tau) + a^2b^2(a-b)\tilde{u}^2(t/\tau)^2 \Big) \tag{4.3}$$

where $a = 2$, $b = N(\tilde{u}+p_0)/\tilde{u}$ and $E_x = e^{-x\tilde{u}t/\tau}$. Note that if the $\tilde{u}$-steps are fast (i.e. if $\tilde{u}t/\tau \gg 1$), the limit of this expression is given by $Z(t) = n_m u(u+p_0)(t/\tau)^2$. In the opposite limit where $\tilde{u}t/\tau \ll 1$, we have $Z(t) = n_m N u(u+p_0)\tilde{u}(\tilde{u}+p_0)(t/\tau)^4/6$.

The key idea of this analysis is to identify how many slow (rate-limiting) steps separate the initial state (X_0) from the state of interest. The slow steps in our model are the ones whose rates scale with u or p_0. The step from Y_1 to Y_2 is fast, because it is proportional to the rate of LOH in CIN cells, p, which is much larger than u and p_0. The steps with the rate $\tilde{u}$ are neither fast nor slow. For all possible pathways from the initial state to the final state of interest, we have to multiply the slow rates together times the appropriate power of t/τ, and divide by the factorial of the number of slow steps. Summing over all possible paths we will obtain the probability to find the crypt in the state in question.

Applying this rule, we can see that $X_2(t)$ and $Y_2(t)$ are both quadratic in time, because it takes two rate-limiting steps to go from X_0 to X_2 and from X_0 to Y_2. The state Z_2 is separated from X_0 by two rate-limiting steps and two 'intermediate' steps (whose rate is proportional to $\tilde{u}$), so the quantity $Z_2(t)$ grows as the forth power of time for $\tilde{u}t/\tau \ll 1$ and as the second power of time in the opposite limit.

The probability that a crypt is dysplastic at time t is given by $P(t) = X_2(t) + Y_2(t) + Z_2(t)$. Therefore, the expected number of dysplastic crypts in a person of age t is $MP(t)$. Of these dysplastic crypts, $MY_2(t)$ have CIN and $MZ_2(t)$ have MSI. This suggests that the fraction of CIN cancers is at least $Y_2(t)/P(t)$ and the fraction of MSI cancers is at least $Z_2(t)/P(t)$. The actual values may be higher because in our model, only the very first stage of cancer development is considered. At later stages of progression from a dysplastic crypt to cancer, there are more chances for cells to acquire a CIN or an MSI mutation.

Some numerical examples are given in Table 4.4, where the relative fractions of dysplastic crypts with CIN, MSI and without genetic instability are presented for different values of n_c, the number of CIN genes. Larger values of n_c lead to increased percentage of dysplastic crypts with CIN. According to observations, 13% of all sporadic colorectal cancers have MSI and 87% have CIN [Lengauer *et al.* (1998)]. In terms of our model this means that we should have $Z_2(t)/P(t) < 0.13$ and $Y_2/P(t) < 0.87$. From Table 4.4 we can see that for values of n_c of the order of 100, the fraction of CIN crypts is higher than expected.

Table 4.4 **Sporadic colorectal cancer**: the expected number of dysplastic crypts and fractions of crypts with different instabilities, at 70 years of age, in the model with CIN and MSI. $M = 10^7$, $N = 5$, $\tau = 20$ days, $u = 10^{-7}$, $u_{met} = 10^{-6}$, $\tilde{u} = 10^{-4}$, $p_0 = 10^{-7}$, $n_m = 3$ and $t = 70$ years.

MSI gene inactivation	n_c	Total number of dyspl. crypts	% of CIN	% of MSI
mutation	1	2	28%	0.4%
mutation	10	8	80%	0.1%
mutation	100	67	98%	0.01%
methylation	1	3	21%	23%
methylation	10	9	74%	7%
methylation	100	68	97%	1%

The calculations presented in Table 4.4 were performed under the assumption that chromosomal instability does not have a cost. In other words, the CIN phenotype is neutral with respect to the wild type. Perhaps a more realistic model should include a possibility that CIN phenotype is disadvantageous compared to the wild-type. Indeed, genetically unstable cells can have a higher apoptosis rate because of a high frequency of mutations in essential genes. Therefore, we can assume that genetic instability leads to a change of reproductive rate thus giving the mutant cells a selective disadvantage, see Chapter 6 (or a selective advantage, if the environment is right, see Chapter 7).

For both disadvantageous and advantageous CIN, we have the computational machinery developed in Chapter 3 which can be used. Let us suppose that the relative reproductive rate of CIN cells is r_c. Then, if no tunneling occurs, the transition rates from X_0 to Y_0 and from X_1 to Y_1 is not $2n_cu$ but $2n_cNr_c^{N-1}(1-r_c)/(1-r_c^N)$. This quantity is larger than $2n_cu$ in the case when CIN is advantageous ($r_c > 1$) and smaller if it is disadvantageous ($r_c < 1$). In the case where CIN is disadvantageous, the transition rate from X_0 to Y_0 and from X_1 to Y_1 becomes lower. For example, if the relative disadvantage of a CIN cell is 10%, then the fraction of CIN dysplastic crypts in Table 4.4 will be reduced by 20%. In the case where stochastic tunneling occurs, the computation will be different. It follows in a straightforward manner from the results of Chapter 3, and we do not present it here. An interested reader can refer to the paper by Komarova et al [Komarova *et al.* (2003)] for details.

In our model, we assume that CIN is generated by means of a mutation in any of n_c dominant-negative CIN genes. In other words, a genetic hit in either of the two copies of a CIN gene will lead to the acquisition of the CIN phenotype. Alternatively, it could happen that the CIN phenotype requires the inactivation of both copies [Rajagopalan *et al.* (2004)], like MSI genes or tumor suppressor genes. In terms of the diagram in Figure 4.3, this would mean that we have two steps separating the wild type (X_0) from the CIN phenotype (Z_0). If we assume that the CIN phenotype is neutral, the fraction of CIN dysplastic crypts in Table 4.4 would be negligible. This means that in this case, the CIN phenotype must be very advantageous in order to show up early in carcinogenesis. We will develop these ideas further in Chapters 6 and 7.

The fraction of MSI crypts as predicted by this model is quite low (for $n_c = 10$ we get only 0.1 % of dysplastic crypts with MSI). This could mean that MSI develops at later stages of cancer. However, there is indirect evidence that the replication error phenotype precedes, and is responsible for, APC mutations in MSI cancers [Huang *et al.* (1996)]. Our model is consistent with this data if we assume higher rates of MSI induction in a cell. This could be caused by higher mutation rates in MSI genes, a larger number of MSI genes or the possibility of epigenetic mechanisms of gene silencing. DNA methylation of the hMLH1 gene is found at a high frequency in sporadic MSI tumors [Ahuja *et al.* (1997); Cunningham *et al.* (1998); Kane *et al.* (1997)]. In the diagram of Figure 4.3 this means that the rates from X_0 to the MSI type (vertical arrows), $2n_mu$ and $u + p_0$, should be replaced by $2n_m u_{met}$ and $u_{met} + p_0$, respectively, where u_{met} is the rate of methylation per gene per cell division. In terms of our equations, we need to replace u by u_{met} in the expression for $Z_2(t)$, equation (4.3). If we assume that u_{met} is larger than the basic mutation rate, u (say $u_{met} = 10^{-6}$), then the expected fraction of MSI crypts predicted by our model becomes larger. Note however that at this stage there is no accurate estimation of methylation rates compared to mutation rates. Our model suggests that if epigenetic mechanisms significantly increase the APC inactivation rate, then the predicted fraction of MSI crypts is consistent with the observed frequency of MSI cancers, see Table 4.4.

4.4 FAP

The framework developed in this chapter allows us to study familial cancers in a systematic manner, by modifying the basic mutation-selection diagram of Figure 4.3. In FAP patients, one allele of the APC gene is inactivated in the germ line. In terms of our model this means that all crypts start in state X_1. The corresponding mutation-selection network is found in Figure 4.4a.

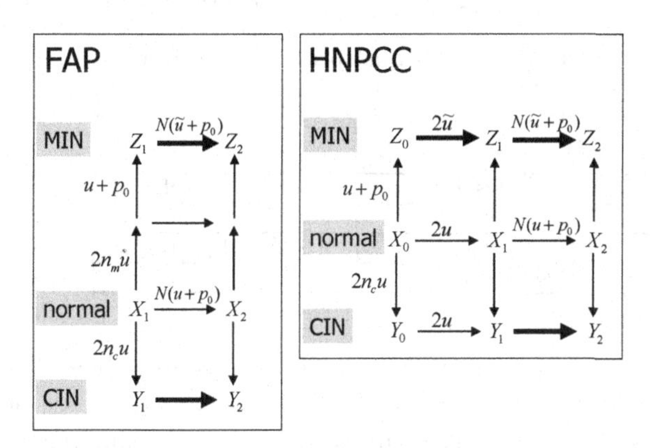

Fig. 4.4 (a) Mutation-selection network of FAP initiation. We start with the type X_1 because the first copy of the APC gene is inactivated in the germ line. (b) Mutation-selection network of HNPCC initiation. One mutation of an MSI gene is inherited, and therefore it takes only *one* step (inactivation of the second copy of the MSI gene, arrows up) to develop the MSI phenotype.

Again, the mutation-selection diagram can be converted into a system of ODEs. The solutions are given by

$$X_2(t) = N(u + p_0)t/\tau, \qquad Y_2(t) = 2n_c ut/\tau,$$

and

$$Z_2(t) = \frac{n_m u(u + p_0)[2 - 2\tilde{u}bt/\tau + (\tilde{u}bt/\tau)^2 - 2E_b]}{(\tilde{u}b)^2}.$$

In the limit where $\tilde{u}t/\tau \to \infty$, we have $Z_2(t) = n_m u(u + p_0)(t/\tau)^2$. If $\tilde{u}t/\tau \ll 1$, then $Z_2(t) = n_m N u(u + p_0)(\tilde{u} + p_0)(t/\tau)^3/3$. $X_2(t)$ and $Y_2(t)$ are linear functions of time (there is one rate-limiting step), whereas $Z_2(t)$ grows slower than the second power of time (two rate-limiting steps plus one 'intermediate' step).

Some predictions of the model are shown in Table 4.5. The expected number of dysplastic crypts and the fraction of CIN crypts are calculated for $t = 16$ years. As the number of CIN genes, n_c, increases, we expect more dysplastic crypts, and a larger fraction of crypts with CIN. According to our model, the expected number of dysplastic crypts grows linearly with time, and by the age of 16 years is expected to be in the thousands to tens of thousands, see Table 4.5. This should be compared with the observation that patients with FAP have hundreds to thousands of polyps by age 16.

Table 4.5 **FAP**: the expected number of dysplastic crypts and the fraction of CIN crypts, at 16 years of age. $M = 10^7$, $N = 5$, $\tau = 20$ days, $u = 10^{-7}$, $\tilde{u} = 10^{-4}$, $p_0 = 10^{-7}$, $n_m = 3$ and $t = 16$ years.

n_c	Total No of dyspl. crypts	% of CIN	% of MSI
1	$\sim 3,500$	17%	0%
10	$\sim 8,800$	67%	0%
100	$\sim 61,300$	95%	0%

The number of polyps in FAP patients does not grow linearly with time. Instead, most polyps appear 'suddenly' in the second decade of life. These observations are consistent with the predictions of our model. It is believed that polyps result from dysplastic crypts by means of further somatic mutations and clonal expansions. Therefore, the number of polyps is expected to be a higher than linear power of time, which looks like a steep increase in the number of lesions after a relatively non-eventful period. Also, the number of dysplastic crypts ($10^3 - 10^4$ in our model) is expected to be much larger than the number of polyps ($10^2 - 10^3$) consistent with the expectation that not all dysplastic crypts progress to polyps.

Another prediction of this model is that the fraction of MSI crypts in patients with FAP is negligible. This is consistent with an experimental study where MSI was found in none of the 57 adenomas from FAP patients [Keller *et al.* (2001)].

Finally, we note that the logical possibility exists that the second copy of the APC gene in FAP patients may be inactivated by an epigenetic event, just like the second copy of an MSI gene can be silenced by methylation. Experimental investigations [Menigatti *et al.* (2001)] however suggest that this is unlikely: out of the 84 FAP tumors, only 1 exhibited hypermethylation of the APC gene.

4.5 HNPCC

Patients with HNPCC inherit one mutation in an MSI gene. The corresponding mutation-selection network is presented in Figure 4.4b. The solutions for X_2 and Y_2 in this case are identical to those for sporadic colorectal cancers and are given by equations (4.2). The solution for Z_2 is as follows:

$$Z_2(t) = \frac{(u+p_0)[a^2E_b - b^2E_a + (a-b)(\tilde{u}abt/\tau - (a+b))]}{ab\tilde{u}(a-b)};$$

in the limit where $\tilde{u}$ is a fast rate we have $Z_2(t) = (u+p_0)t/\tau$. In the opposite limit, where $\tilde{u}$ is a slow rate, $Z_2(t) = N(u+p_0)\tilde{u}(\tilde{u}+p_0)(t/\tau)^3/3$. If we assume that the second copy of the MSI gene is silenced by methylation, we need to replace u by u_{met} in the expression for $Z_2(t)$.

The solutions for X_2 and Y_2 in this case are quadratic in time (two rate-limiting steps), and the quantity $Z_2(t)$ grows slower than linear but faster than quadratic, because it requires one rate-limiting and two intermediate steps (note that we are talking about linear and quadratic functions of an argument smaller than one). In Table 4.6 we present the expected number of dysplastic crypts and the fraction of MSI crypts, calculated for $t = 40$. We have explored two possibilities: (1) inactivation of the second copy of an MSI gene happens by means of a point mutation, with the rate u, and (2) inactivation of the second copy of an MSI gene happens by methylation. There is evidence that the second scenario is less likely in the case of HNPCC [Yamamoto *et al.* (2002)]. In a recent study, DNA methylation of the hMLH1 gene was found in 80% of 40 sporadic MSI cancers and in 0% of 30 cancers in HNPCC patients [Esteller *et al.* (2001)].

Our model predicts that the majority of dysplastic crypts in HNPCC patients are expected to have MSI. However, we do not find that 100% of dysplastic crypts will contain MSI. On the other hand, we know that virtually all tumors in HNPCC patients have MSI. This might suggest that selection for MSI also happens at later stages of carcinogenesis: dysplastic crypts with MSI might have a faster rate of progression to cancer than dysplastic crypts containing CIN or normal cells.

Finally we note that the total number of dysplastic crypts in HNPCC patients, as predicted by our model, is of the order 10 at age 40, which is only slightly larger than the expected number of dysplastic crypts in normal individuals and is not nearly as high as in the case of FAP (of the order 10,000, Table 4.6). This is also consistent with observations.

Table 4.6 **HNPCC**: the expected number of dysplastic crypts and the fraction of MSI crypts, at 40 years of age. $M = 10^7$, $N = 5$, $\tau = 20$ days, $u = 10^{-7}$, $u_{met} = 10^{-6}$, $\tilde{u} = 10^{-4}$, $p_0 = 10^{-7}$, $n_c = 10$ and $t = 40$ years. Compared with patients with FAP and sporadic colorectal cancer.

Condition	Total No of dyspl. crypts	% of CIN	% of MSI
HNPCCby mutation	14	15%	81%
HNPCC by methylation	66	3%	96%
Sporadic colorectal cancer	3	78%	3%
FAP	$\sim 22,000$	67%	0%

4.6 Insights following from this analysis

In this chapter we applied the tools developed in Chapter 3 to study the dynamics of colorectal cancer initiation. We calculate the rate of dysplastic crypt formation as a consequence of inactivating both alleles of the APC tumor suppressor gene. This can either happen in normal cells or in cells that have already acquired one of the two genetic instabilities, MSI or CIN. If the rate of triggering genetic instability in a cell is high and if the cost of genetic instability is not too large, then inactivation of APC will frequently occur in cells that are genetically unstable. In this case, genetic instability is the first phenotypic modification of a cell on the way to cancer.

It is interesting to compare the two types of instability, MSI and CIN. MSI, being associated with subtle changes in the genome, is probably less of a liability for the cell than CIN. In other words, CIN cells are more likely to produce non-viable offspring than MSI cells. At the same time, it may be possible that CIN is easier to trigger (for instance, if it requires a change in a single allele of many genes). Our analysis shows that if inactivation of MSI genes (either by point mutation or by methylation) occurs at a sufficiently fast rate - around 10^{-6} per cell division, then MSI can precede APC inactivation in a significant number of cases. Regarding CIN, the crucial questions are (i) how many dominant CIN genes can be found in the human genome, (ii) how fast are CIN genes inactivated, and (iii) what are the costs of CIN. A more detailed analysis of costs and benefits of CIN is given in Chapters 6 and 7.

Our calculations show that important insights could be derived by carefully monitoring the incidence rate of dysplastic crypts in patients as function of age. With or without early genetic instability, the abundance of dysplastic crypts should grow approximately as a second power of time. The two rate limiting steps can either refer to two mutations of APC, or one mutation of APC and one CIN mutation. In the case of CIN, LOH of the second allele of APC is not rate limiting. Hence, two rate limiting steps for the inactivation of a tumor suppressor gene can be compatible with an additional genetic instability mutation.

Several further insights emerge from our analysis.

Fraction of dysplastic crypts with CIN or MSI. About 87% of sporadic colorectal cancers have CIN while the rest have MSI. Assuming that CIN and MSI are irreversible, we conclude that the maximum fraction of dysplastic crypts with CIN should be 87%, while the maximum fraction of dysplastic crypts with MSI should be 13%. This provides certain restrictions on the possible parameter values of our model (see Table 2b).

Epigenetic factors. If we assume that MSI genes in sporadic colorectal cancer are inactivated only by point mutation or LOH events, then the fraction of dysplastic crypts with MSI is very low. We get higher fractions of MSI if we assume that MSI genes can also be inactivated by methylation and if methylation of MSI genes is fast compared to point mutation or LOH. Thus, methylation events could play a crucial role in the formation of sporadic MSI cancers.

Competition among crypts. Another interesting possibility is that dysplastic crypts can be lost and replaced by normal crypts. In this case, many dysplastic crypts could be produced, but only a part of them is retained so that the actual number of dysplastic crypts stays low. To our knowledge, the competitive dynamics of crypts in a colon has not been investigated experimentally.

No MSI in FAP. Our model predicts that the fraction of MSI dysplastic crypts in FAP patients is close to zero. A significant number of dysplastic crypts will contain CIN. This is consistent with experiments observations.

The number of dysplastic crypts. We calculated both the absolute numbers and relative proportions of dysplastic crypts with or without genetic instabilities. An interesting empirical project is to measure the abundance of such dysplastic crypts as function of age. This will provide crucial information on the dynamics of colorectal cancer initiation.

A more precise description of the mutation spectrum. The mutation spectrum of the APC gene is far from random (one reason being that the APC gene is long and multi-functional). The type of the second APC mutation may depend on where the first APC mutation took place [Lamlum *et al.* (1999); Rowan *et al.* (2000)]. Our model is well suited to take this into account. Here is a simple way to differentiate between two kinds of point mutations. Let us assume that the total probability of a point mutation is u (as in the basic model), and there are two kinds of mutations. (i) With probability u_1, a mutation happens such that the second allele can *only* be inactivated by a point mutation. (ii) With probability u_2, a mutation happens which can be followed by another point mutation *or* an LOH event. We have $u_1 + u_2 = u$. These two scenarios can be incorporated in our calculations adding a new level of complexity to the basic theory.

The cells at risk of cancer. In this first model we assumed that only stem cells are at risk of cancer. Another possibility is that both stem cells and large numbers of differentiated cells in a crypt are running the risk of acquiring cancerous mutations. In its present form, this analysis would predict that the expected number of dysplastic crypts in persons of 70 years of age is enormous and biologically implausible. In order to correctly include the possibility of cancer initiation in partially differentiated cells, one needs to perform a calculation similar to that presented in the next chapter.

Chapter 5

Cellular origins of cancer

Chapter 3 presented an extensive stochastic analysis of a two-hit model. In particular we calculated the probability of creating a double-mutant as a function of time, depending on the population size and the relative fitness of the intermediate type. Chapter 4 made the first attempt to apply this model to real-life carcinogenesis, by taking account of specific features of sporadic and familial colorectal cancers. One important consideration which was not included in the analysis so far is the population structure. In Chapters 3 and 4, the population of cells was completely homogeneous with respect to the patterns of mitosis/apoptosis. In other words, cells were only characterized by their "fitness", which was a function of acquired mutations. In some cases, this is not enough to grasp the essential dynamics of the system. An example is the colonic epithelial tissue. There, when talking about the dynamics of cell division and mutations, we may have to take into account the fact that stem cells behave differently from differentiated cells. The analysis which follows can be of importance for one of the fundamental questions in cancer research, namely, *from which cells in our body does cancer originate?*

To begin, we will briefly describe important aspects of tissue architecture, development, and function. We have to make a distinction between *stem cells* and *differentiated cells.* Stem cells have the ability to divide indefinitely. During this process they give rise to differentiated cells which make up the tissue. The differentiated cells perform their function and eventually die. In the context of stem cells, we have to distinguish between *embryonic stem cells* and *adult stem cells.* Embryonic stem cells give rise to the organism during development. They are said to be truly multipotent. That is, they can give rise to any tissue in the body (e.g. lung, liver, brain, colon, skin, etc.). Adult stem cells, on the other hand, are thought to be more restricted. That is, they might only be able to give rise to certain tissues. For example, liver stem cells can only differentiate into "committed" liver cells, or colon stem cells can only differentiate into "committed" colon cells. Adult stem cells are thought to be responsible for maintaining and renewing a given tissue. They may divide at a relatively slow rate, or divide only when new tissue cells need to be created (e.g. when already differentiated cells die). Division of adult stem cells is thought to be *asymmetric*. That is, division gives rise to one stem cell, and one cell which differentiates into a functioning tissue cell.

Consider the colon as an example. The epithelial lining of the colon is made up of many involutions which are called *crypts*. There are about 10^7 crypts within a human colon. Each crypt contains stem cells. The exact number of stem cells per crypt is not known; there might be just one stem cell or a small number of them. Upon division, a stem cell gives rise to one stem cell and one cell which embarks on a journey of differentiation. Before this cell is fully differentiated, it divides a certain number of times. A fully differentiated cell lives for about one week. Then it dies and is washed out of the colon. The first malignant change in colon cancer ensures that the differentiated cell does not die after one week. Instead it remains, and this causes an accumulation of abnormal or transformed cells. The inactivation of the tumor suppressor gene APC is responsible for this behavior. The generation of APC-/- cells (or the inactivation of other genes involved in the Wnt pathway) is the first step toward colon cancer [Katoh (2003); Kinzler and Vogelstein (1998); Polakis (1997); Polakis (1999)]. We are faced with an important question. Did the mutation which inactivates the APC gene occur in the stem cells, or in the cells which differentiate?

Talking in more general terms, cancer cells have been shown to have various characteristics in common with stem cells. Fore example, they have the capacity to divide indefinitely. This does not, however, mean that the origin of cancer lies in stem cells. There are two theories. The *stem-cell theory*, suggests that the first event happens in a stem cell. The *de-differentiation* theory, claims that it occurs in a (partially)-differentiated cell, thus leading to its de-differentiation, or "immortalization". Experimentally this is a very difficult question, and the debate is ongoing.

Despite this uncertainty, many researches feel that differentiated cells are unimportant for cancer initiation, for the following (quantitative) reason, unrelated to biological evidence. Let us concentrate on colon cancer. It is widely believed that the APC gene is a tumor suppressor gene. That is, the inactivation of both copies is required to confer phenotypic changes

[Kinzler and Vogelstein (1998); Macleod (2000)] (consistent with Knudson's two-hit hypothesis [Knudson (1996)]). Then an obvious question arises: how can the first mutation occur in the migrating compartment, without being washed away? As John Cairns writes [Cairns (2002)], "...there are 256 exponentially multiplying cells that divide twice a day and are being replenished continually by the divisions of a single stem cell, none of these 256 cells will ever be separated from the stem cell by more than eight divisions, and the replication errors made in those eight divisions are destined, of course, to be discarded".

The point of this chapter is to address exactly this issue: will migrating cell mutations be indeed discarded, or is there a chance that they will persist until the second hit comes, which immortalizes the cell and thus initiates dysplasia in the crypt?

5.1 Stem cells, tissue renewal and cancer

The normal functioning of colon relies on the fine-tuned balance of the epithelial cell production, differentiation and death. The regulation of the processes of cell proliferation and shedding occurs at the level of crypts - the folds of colonic epithelium which are continuously renewed by stem cell division. The appearance of dysplastic crypts in the beginning of colorectal cancers is a manifestation of the broken balance between cell division and apoptosis. At the molecular level, it has been shown that the earliest event of sporadic colorectal cancers is the inactivation of the APC gene [Kinzler and Vogelstein (1998); Polakis (1997)], or other genes involved in the Wnt pathway [Katoh (2003); Polakis (1999)]. The APC gene inhibits members of the Wnt signaling pathway, which promote the expression of β-catenin. In its turn, β-catenin acts as an enhancer of cell division [Behrens *et al.* (1998)].

It is widely believed that the relevant target cells for the first mutation are the colonic stem cells, [Bach *et al.* (2000); Fuchs and Segre (2000); Kim and Shibata (2002); Potten *et al.* (2003); Potten and Loeffler (1990); Winton (2001); Wong *et al.* (2002)]. The argument usually goes in the following way, see e.g. [Cairns (2002)]. If the first mutation happened in a proliferative daughter cell, it would be washed away before the second hit has a chance to confer a significant phenotypic change. On the other hand, if the first mutation occurs in the "immortal" stem cell, then its mutant progeny will populate the compartment and persist for as long as it takes to accumulate further mutations which give rise to neoplasia.

It is often assumed that the stem cells are located in a niche at the base of the crypt. The "bottom-up" model of colorectal histogenesis [Preston *et al.* (2003); Wong *et al.* (2002); Wright and Poulsom (2002)] states that the oncogenic mutations occur in the stem cells at the base of the crypt. A natural consequence of this model is that once such a mutation occurs, the entire crypt will be monoclonally-mutant. There has been some evidence which contradicts this view. Namely, dysplastic cells exhibiting genetic alterations in the APC gene have been found in the upper layer of crypts, whereas cells located at the bottom of the same crypt did not contain such alterations. This led to the "top-down" model [Shih *et al.* (2001a)]. The two explanations proposed were that (i) the stem cells reside near the top of the crypts, or (ii) transformed cells originate from the stem cells at the base of the crypt, then they passively move upward, after which a cycle of cell proliferation and tissue replacement starts in the top-down direction.

Another explanation has been put forward which suggests that the relevant mutations occur in fully differentiated cells [Fodde *et al.* (2001b)], or in the proliferative/migrating daughter cells [Lamprecht and Lipkin (2002)]. This is consistent with the stem cells being located at the base of the crypt, as previously thought. At the basis of these models is the idea that a proliferating daughter cell with a silenced APC gene (or otherwise upregulated β-catenin) will acquire a stem cell phenotypic characteristic, that is, permanence in the crypt. Indeed, it has been shown that the β-catenin/T cell factor 4 complex constitutes the "master switch" that controls proliferation versus differentiation in healthy and malignant intestinal epithelial cells [van de Wetering *et al.* (2002)]. Inactivation of the APC gene in a migrating cell could reverse the process of differentiation and trick the cell into thinking that it is "immortal". This would lead to a continued proliferation of this cell which would avoid entering the final differentiation and programmed apoptosis stage.

The following sections will discuss these mechanisms mathematically.

5.2 The basic renewal model

We assume that the stem cells are located in a niche at the base of the crypt. They are characterized by an asymmetric division pattern resulting in one stem cell and one proliferative daughter cell. The latter cells divide and populate the migrating compartment. Cells of the migrating compartment go through a number of symmetric divisions, moving toward the crypt surface. On their way up, they go through stages of differentiation, until the fully differentiated cells stop dividing, reach the crypt surface and get shed into the lumen, to be replaced by new generations of cells coming from the bottom of the crypt. We will refer to the symmetrically dividing, migrating progeny of a stem cell (SC) as "differentiated cells" (DC), keeping in mind that the degree of differentiation increases with the number of divisions that separate the progeny from the stem cell.

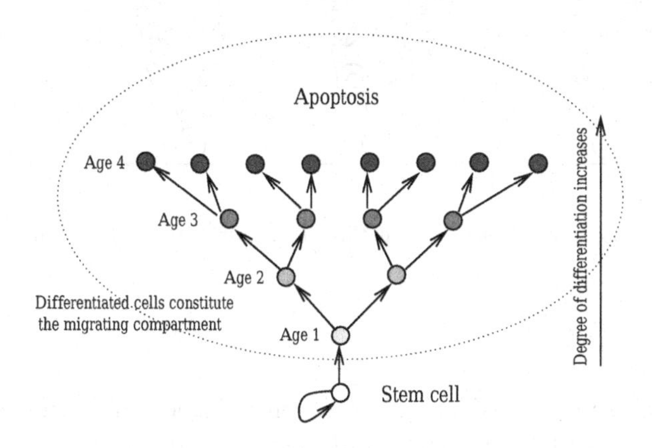

Fig. 5.1 The history of one daughter DC. It undergoes 3 rounds of division. The number of cells in the last generation is 8.

Figure 5.1 traces the offspring of one DC created from a SC. Different levels represent consecutive moments of time (or rounds of proliferation). Another interpretation of this figure is spatial: we can think of cells of consecutive generations to be located closer and closer to the top of the crypt. The cells of "age 4" are the closest to the top, and they are shed into the lumen. This can be better seen in Figure 5.2.

The progeny of a single daughter cell is marked by the same letter. The degree of maturation/differentiation is reflected in the intensity of shading: the darker the circle is, the more mature is the cell. The apoptotic cells are presented by dashed circles. At all moments of time, the crypt contains DCs of 4 generations. In the beginning (the leftmost diagram) there are progeny of cells A, B and C, and a newly produced daughter cell, D. After some time (the middle diagram), all of the "oldest" cells (marked with A) have been shed, the cells B and C went through one round of division, advancing their degree of maturation and moving upward, and a daughter

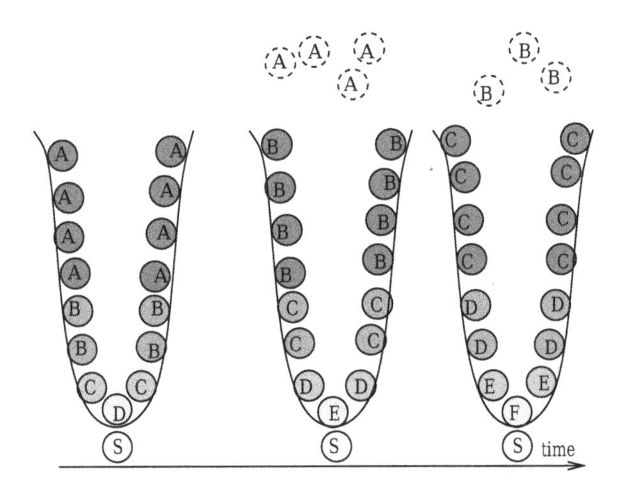

Fig. 5.2 Schematic snapshots of one crypt at three moments of time. The stem cell is marked by "S".

cell, E, has been produced. The process of renewal goes on in this way, eventually replacing all DCs in the crypt [Komarova and Wang (2004)].

In Figure 5.2 only one stem cell is shown to repopulate the crypt. In reality, there are several stem cells per crypt, and thus each crypt is a composition of several clones. This can be easily included in our model: if, for example, there are four stem cells in the crypt, then this crypt can be viewed as a "superposition" of four crypts. The general rule that more mature cells are situated closer to the top of the crypt still holds, and the general dynamics of each of the clones is as in Figure 5.2. The only difference is in the numbers that should be used in the model.

Several models of SC dynamics have been designed, [Ro and Rannala (2001); Yatabe *et al.* (2001)]. In these papers, two main mechanisms of SC reproduction have been proposed. In the deterministic model, each SC divides asymmetrically, and the number of SC is kept constant. In more sophisticated models, each SC has a probability to produce upon division (i) two SCs, (ii) one SC and one DC, or (iii) two DCs. In this model, the number of stem cells fluctuates. The latter model seems to be more realistic. Here, we will use the simpler model, and note that the methods and results developed should remain the same, with minimal changes, if the reproduction model for stem cells is refined.

5.3 Three scenarios

Let us first describe the process of accumulation and spread of mutations in a dynamic crypt. By "mutations" we mean any kind of genetic alteration (a point mutation, a loss of heterozygocity (LOH) event, etc.) which leads to the inactivation of an allele of the APC gene. We will assume that a cell with a single mutation has the same properties as a wild type cell, and a double-mutant has the ability to avoid apoptosis, continue divisions and thus remain and spread in the crypt.

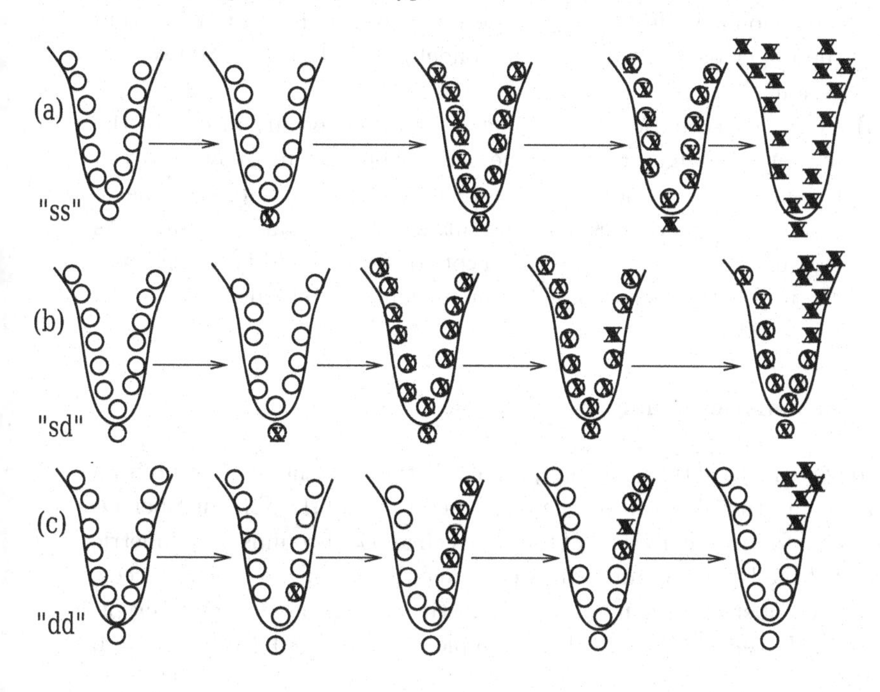

Fig. 5.3 Three scenarios of the emergence of a double mutant. (a), the *ss* scenario. (b), the *sd* scenario. (c), the *dd* scenario. The open circles are wild-type cells, circles with an "x" contain one mutation in the tumor suppressor gene, and double-x's are double mutants.

There are three logical possibilities of accumulation of mutations, see Figure 5.3.

(i) In the *ss* scenario, Figure 5.3a, a mutation happens in the stem cell. Then, after a few divisions, the entire crypt will consist of mutated cells. At some point, a *second* mutation occurs in the SC, shortly after which the entire crypt will consist of double mutants. This is the scenario consistent with the "bottom-up" hypothesis. It predicts that the crypt will be monoclonal with respect to double mutations.

(ii) In the *sd* scenario, Figure 5.3b, again a mutation occurs in the SC which then spreads throughout the crypt. However, the first double-mutant emerges in the proliferating/migrating compartment. This mutant divides and its progeny first spreads in the upward direction. At this point, the lower part of the crypt is monoclonal with respect to one mutation in the APC gene, and the upper part of the crypt is monoclonal with respect to two mutations.

(iii) In the *dd* scenario, Figure 5.3c, a mutation occurs in one of the migrating daughter cells. The cell divides, its progeny moves in the upward direction, but before it undergoes apoptosis, one of these cells experiences a second hit, creating a double mutant. As a result, the lower part of the crypt consists of wild-type cells, and the upper part is composed of monoclonal double-mutants.

5.4 Mathematical analysis

It is convenient to introduce the quantity l, the total number of division rounds during the life-span of one clone, see Figure 5.1. This number includes one asymmetric division of the stem cell and $l-1$ rounds of symmetric divisions of the DCs (for simplicity we assume that these are synchronized). The total number of progeny of an SC existing at any one time in a crypt, is $2^l - 1$. We denote $N = 2^l$. For example, in Figure 5.1 we have $l = 4$, $N = 16$.

Since the probability of the first hit, p_1, is very small, it can be shown that most of the time the population of the crypt will be homogeneous, that is, most of the cells will either be wild-type, or will contain one mutation, see also [Komarova *et al.* (2003)]. Indeed, if a mutation arises in a DC, it gets washed out in less than l time-steps; in fact, most of the mutants are very short-lived, and only survive for one or two time-steps, because at each moment of time, the majority of the crypt consists of cells only one or two steps away from apoptosis, see Figure 5.1. The frequency with which new mutants are created is Np_1, and only about $1/N$ of these mutants will happen in an SC. So the condition $p_1 \ll 1/N$ (the *homogeneity* condition) guarantees that the crypt contains no mutants most of the time. Unless, of course, a mutation occurs in a SC, in which case in less than l time-units, the entire crypt will consist of mutant cells. Finally, a double mutant may appear, but in this case we assume that the process is over and a dysplastic crypt has been created. Note that the homogeneity condition easily holds for the realistic values, $p_1 \sim 10^{-7}$ [Albertini *et al.* (1990)], and $N \sim 10^3$ [Kim and Shibata (2002); Potten and Loeffler (1990)].

Let us call the probability to find the entire crypt consisting of wild-type cells, x_0. The probability that the entire crypt consists of cells with a single mutation is x_1, and the probability that the crypt is dysplastic (contains one or more double mutants) is x_2. The symbols X_0, X_1 and X_2 will be used to denote the corresponding states. Because of the homogeneity condition, we have $x_0 + x_1 + x_2 \approx 1$. A double mutant can be created via two major pathways. One pathway includes a fixation of a single mutant. First, a mutation happens in a SC (the rate is p_1), after which the entire crypt enters state X_1, and then a new mutation occurs (with rate Np_2), which brings the crypt to the state X_2. Alternatively, a second mutation can occur in a mutant clone which originates in a DC, without a prior fixation of a single mutant. This happens with the rate R, which we now calculate.

Calculation the rate of double mutant generation, for path *dd*. Let us write down the probability to have at least one mutant such that both mutations happen in the DC (given that no SC mutations have happened). We assume that the mutation rate, p_1, is sufficiently small such that the clones can be treated independently (the condition is $p_1 N \ll 1$), and consider a doubly stochastic process, see also [Iwasa *et al.* (2004); Moolgavkar *et al.* (1988)]. We obtain

$$R = \sum_{i=1}^{l} r_i \mathcal{P}_i.$$

To see this, we write down the total rate of primary mutations, $\sum_{i=1}^{l} p_1 2^{i-1} = p_1 N$ (this is valid for $p_1 N \ll 1$). Then each term $r_i = p_1 2^{i-1}$ is a contribution corresponding to the first mutation happening in a DC of generation i. $\mathcal{P}_i$ is related to the "secondary" stochastic process which happens in the clone after the first mutation. Inside such a clone, we have $2^{l-i+1} - 2$ cell divisions, so that the total probability to get a second hit is

$$\mathcal{P}_i = 1 - (1 - p_2)^{2^{l-i+1}-2} \approx 1 - e^{-p_2(2^{l-i+1}-2)}.$$

We have

$$R = p_1 \sum_{i=1}^{l} 2^{i-1}(1 - e^{-p_2(2^{l-i+1}-2)}).$$

This expression can be calculated by replacing the summation with an integral,

$$R = \frac{Np_1}{\log 2}\left[\frac{1}{2} - \frac{1}{N} - e^{2p_2}\int_{\frac{1}{N}}^{\frac{1}{2}} e^{-p_2/x}\,dx\right].$$

In the limit of small p_2, we have the following expression,

$$R \approx \frac{Np_1p_2(|\log p_2| - \log 2 - \gamma)}{\log 2} \tag{5.1}$$

(here γ is the Euler's constant, $\gamma \approx 0.577$).

It is interesting to compare this with the results of Chapter 3, obtained for the rate at which a double mutant is produced in a two-hit model. In Chapter 3, the structure of compartments is not taken into account, but the phenomenon of "stochastic tunneling" tunneling that we introduced there is very similar to pathway *dd* studied here. Indeed, tunneling occurs when the second hit occurs before the first hit has had a chance to reach fixation. For such models, the rate R, roughly speaking, is given by $Np_1\sqrt{p_2}$. We can see that taking account of the structure of the colon changes these results. In particular, the new rate given by (5.1) is lower because $p_2|\log p_2| < \sqrt{p_2}$.

Equations containing all scenarios. We have the following equations for the mutation processes:

$$\dot{x}_0 = -p_1x_0 - Rx_0, \tag{5.2}$$

$$\dot{x}_1 = p_1x_0 - Np_2x_1, \tag{5.3}$$

$$\dot{x}_2 = Np_2x_1 + Rx_0. \tag{5.4}$$

The probability to obtain a double mutant, x_2, is given by equation (5.4). It can be rewritten as a sum of three contributions,

$$x_2 = x^{dd} + x^{sd} + x^{ss}.$$

Here, the probability to obtain a double mutant by two DC mutations is given by

$$\dot{x}^{dd} = Rx_0,$$

which yields

$$x^{dd} = \frac{R(1 - e^{-(p_1+R)t})}{p_1 + R}. \tag{5.5}$$

The probability to obtain a double mutant by first mutating the stem cell is

$$\dot{x}^{sd} + \dot{x}^{ss} = Np_2 x_1, \tag{5.6}$$

where x^{ss} refers to the pathway where both mutations occur in the SC, and x^{sd} implies the second mutation in a DC. We have the following intuitive relation,

$$x^{ss}/x^{sd} \approx 1/N. \tag{5.7}$$

This means that the vast majority of double mutants will acquire the second mutation in the differentiated stage. We can solve equations (5.2), (5.3) and (5.6) and use equation (5.7) to eliminate x^{ss} to obtain the following expressions:

$$x^{sd} \approx \frac{p_1 A}{(p_1 + R)(p_1 + R - Np_2)}, \tag{5.8}$$

where $A = Np_2(e^{-(p_1+R)t} - 1) - (p_1 + R)(e^{-Np_2 t} - 1)$, and

$$x^{ss} = x^{sd}/N. \tag{5.9}$$

Relative importance of the three scenarios. It is clear that the probability of the *ss* scenario is small compared to the *sd* scenario. As time increases, we have

$$\lim_{t\to\infty} x^{dd} = \frac{R}{p_1 + R}.$$

This means that if the condition $R > p_1$ is satisfied, then the majority of double mutants will acquire both mutations in a DC. Using expression (5.1), we obtain the following inequality,

$$p_2 |\log p_2| > \frac{1}{N}. \tag{5.10}$$

If this condition is satisfied, then the probability to obtain a double mutant by two hits in the DC compartment is larger than the probability to first get a mutation in an SC. It is interesting that this condition only depends on the second mutation rate, p_2, and is independent on the first mutation rate, p_1.

A technical note: very little elaboration is required to obtain the relationship between x^{ss} and x^{sd}, equation (5.7). This simple relationship provides important insight that "stem cells are not the entire story, and the second hit is more likely to occur in a DC". However, we need to do more work to answer the following question: how likely is it that **both** hits fall outside the stem cell compartment? This is where the above model becomes necessary.

Simulations. To check our analytical results, we can use numerical simulations to study the crypt dynamics. Let us trace the offspring of one stem cell; the process goes on until the first double mutant is created. We record whether each of the hits in the double mutant was the result of an SC or a DC mutation, and then we stop the simulation for the crypt. This process is repeated many times, and then the probability of different pathways is estimated. We ran the simulation with 10^3 and 10^7 realizations. The results are almost the same which suggests convergence.

Figure 5.4 plots the probability of having a double mutant via different pathways by time n. The curves represent analytical prediction by formulas (5.5), (5.8) and (5.9). The points are results of numerical simulation and they agree with the calculations very well. Notice that for $p_1 = p_2 = 10^{-4}$ (Figure 5.4a) the probability of having a double mutant initiated in an SC (pathway *sd*) is greater than the probability of having a double mutant initiated in a DC (pathway *dd*). On the other hand, when p_2 increases to $p_2 = 10^{-3}$, the situation reverses, see Figure 5.4b. The reversal of the two scenarios is predicted correctly by formula (5.10): if $N = 2^9 = 512$, this condition holds for $p_2 = 10^{-3}$ and it does not hold for $p_2 = 10^{-4}$.

Figure 5.4c shows the situation corresponding to colon cancer initiation in the presence of chromosomal instability. The first mutation rate, $p_1 = 10^{-7}$, corresponds to the basic point mutation rate. The second mutation rate corresponds to a highly elevated rate of LOH in unstable cancers, $p_2 = 10^{-2}$. We can see that the *dd* scenario prevails in this case.

Simulating crypt dynamics for lower values of mutation rates, e.g., $p_1 = p_2 \approx 10^{-7}$, which corresponds to both hits occurring with the basic point-mutation rate, is very time-consuming. More sophisticated numerical methods could give a faster performance, but here we would like to emphasizes the value of analytical results. Our formulas are valid for arbitrarily small values of mutation rates, and in fact, the precision of the method grows as $p_{1,2}$ decrease!

Finally, we have derived an exact analytical formula for the total prob-

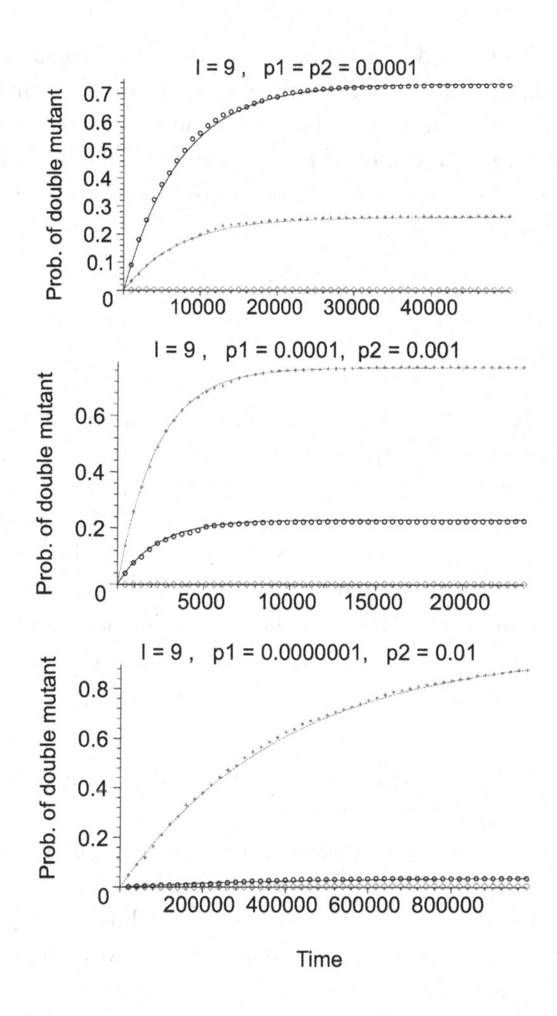

Fig. 5.4 The probability to acquire a double mutant as a function of time, with $l = 9$. Numerical simulations are compared with analytical results. "o" denotes the simulation for *sd* pathway, "+" denotes the simulation for *dd* pathway, and "◇" denotes the simulation for *ss* pathway. In (a), pathway *sd* is more likely. In (b), where p_2 is an order of magnitude larger, pathway *dd* prevails. In (c), *dd* pathway prevails. Pathway *ss* is always the least likely scenario.

ability of creating a double mutant, see Appendix 12.6. This was used to check the simulations by comparing the numerical value of $x^{ss} + x^{sd} + x^{dd}$ with the exact formula, expression (A.1) in Appendix 12.6. The formula gave a perfect agreement with simulation results (data not shown). Formula (A.1) works very well for large values of $p_{1,2}$. For instance, it can give the probability of producing a double mutant, where the approximation of this section breaks down, that is, in the regime where $p_1 \sim 1/N$. However, the exact calculation of Appendix 12.6 has the same disadvantage as the numerical simulations: for low values of $p_{1,2}$ it becomes difficult to implement. The approximate formulas (5.5), (5.8) and (5.9) can be used for all realistic values of mutation rates.

5.5 Implications and data

We considered the renewal dynamics of a colon crypt, repopulated by asymmetric divisions of stem cells. Dysplasia occurs as a result of inactivation of both copies of the APC gene (or other genes in the Wnt pathway). Thus, it takes two hits before phenotypic changes occur. There are three pathways for the two hits: *ss* (both hits occur in an SC), *sd* (the first hit occurs in an SC, and the next hit is acquired by one of the DC in its clone), and *dd* (the first hit is acquired in a DC, and the second hit occurs in one of its progeny). The results obtained from the mathematical models can be summarized as follows:

(1) The probability of the *ss* pathway is negligible. That is, it is unlikely that both hits occur in the SC.
(2) As a consequence, at least one of the hits will occur in the migrating, proliferative compartment. This is consistent with the observation that below the dysplastic layer, cells in the crypt do not exhibit the $\text{APC}^{-/-}$ phenotype. The relative importance of *sd* and *dd* pathways depends on the parameters of the system.
(3) In particular, if $p_2|\log p_2| > 1/N$, the *dd* pathways become more important. It means that the first double mutant will appear *outside* of the SC compartment. This contradicts previous thinking that SCs are crucially important for colon cancer initiation, being the first mutational "targets".
(4) If the reverse condition holds, that is, if $p_2|\log p_2| < 1/N$, then the most likely scenario is the *sd* pathway, that is, first an SC acquires a mutation, then all of its offspring contain an inactivated APC copy, and then one of these daughter cells acquires a second hit. According to the *sd* scenario, the crypt below the dysplastic cells should contain cells with one inactivated and one functional copy of the APC gene.

One remarkable property of the model is a very small number of parameters that it contains. For instance, the condition which identifies whether an SC mutation is important (that is, which of the pathways *sd* and *dd* is more likely), only depends on two parameters, p_2 and N. Table 5.1 identifies the most important pathway, depending on the parameter p_2 and the number of SCs per crypt. The parameter N can be found as follows. If there are n SCs per crypt, then N is the size of the crypt divided by n, that is, the "share" of each of the SCs in the crypt. We suppose that there are 2,000 cells in a crypt, and the number of SCs per crypt is varied between 1 and 32 [Potten and Loeffler (1990)].

Table 5.1 The most likely pathway for the two hits

	Number of stem cells per crypt					
p_2	1	2	4	8	16	32
10^{-7}	*sd*	*sd*	*sd*	*sd*	*sd*	*sd*
10^{-5}	*sd*	*sd*	*sd*	*sd*	*sd*	*sd*
10^{-4}	*dd*	*sd*	*sd*	*sd*	*sd*	*sd*
10^{-3}	*dd*	*dd*	*dd*	*dd*	*sd*	*sd*
10^{-2}	*dd*	*dd*	*dd*	*dd*	*dd*	*dd*

The parameter p_2 is the rate of inactivation of the second allele of the APC gene. In Table 5.1, it is varied between 10^{-7} per cell division (the basic point mutation rate, [Albertini *et al.* (1990)]) and 10^{-2}. The elevated values of p_2 may be related to the phenomenon of chromosomal instability, which is commonly observed in colon cancers [Lengauer *et al.* (1998)]. There is evidence that chromosomal instability arises very early in colorectal cancers [Shih *et al.* (2001b)], and the *in vitro* estimates of the corresponding elevated rate of LOH give the value as high as $p_2 = 10^{-2}$ per cell division [Lengauer *et al.* (1997)]. We can see that without chromosomal instability (low values of mutation rate), the pathway *sd* is more likely. For elevated rates of APC inactivation, it is more likely that both hits will occur in the migrating compartment.

There are several examples in the literature which support the notion that cancer cells might arise in partially differentiated cells. One such example is leukemia, [Grisolano *et al.* (1997)]. Acute myeloid leukemia (AML) is assumed to reflect transformation of a primitive stem cell compartment. On the other hand, it is thought that acute promyelocytic leukemia (APL) arises in committed myeloid progenitors. The evidence in support of the partially-differentiated-cell origin of APL has been reported in [Turhan *et al.* (1995)], where clonality of leukemia was studied. Furthermore, many mouse models of APL have been designed [Brown *et al.* (1997); Grisolano *et al.* (1997); He *et al.* (1997); Westervelt *et al.* (2003); Westervelt and Ley (1999)], which suggest that genetic changes occurring in committed cells may lead to cancer initiation. For some solid tumors, it has also been proposed that the first change may occur in differentiated cells; this includes skin cancers [Perez-Losada and Balmain (2003)], and stomach cancers [Kirchner *et al.* (2001)]. However, for most other cancers, including colorectal cancers, no experiments have so far been performed to clearly establish whether a partially differentiated cell can be a target for cancer initiation.

Chapter 6

Costs and benefits of chromosomal instability

Chapter 4 discussed mathematical models which demonstrated that genetic instability can sometimes speed up the generation of a mutant cell which can give rise to cancer. This underlines the verbal arguments which were first presented in Lawrence Loeb's mutator phenotype hypothesis [Loeb *et al.* (1974)]: the multi-stage nature of cancer initiation and progression requires genetic instability; otherwise a sufficient number of mutations cannot be generated during the life time of a human being.

The mutator phenotype hypothesis, however, considers genetic instability in general and in all its forms. Do the same arguments apply to all types of instabilities? As reviewed in Chapters 1 and 4, genomic instability can be divided into two broad categories: small scale mutations (such as MSI) and gross chromosomal abnormalities (such as CIN). After a mutant cell has been generated, it needs to give rise to clonal expansion for the cancer to be established. If the instability induces the generation of subtle sequence changes, the process of clonal expansion is not likely to be influenced to a significant degree, and the result derived in Chapter 5 still holds. If the instability induces destructive genomic changes, such as imbalances in genes and chromosome numbers, then clonal expansion can be compromised significantly. Although the cells can undergo uncontrolled growth, genome destruction can result in frequent cell death and this could counteract the establishment of a cancer. Therefore, while CIN can speed up the generation of a cell with an inactivated tumor suppressor gene, it might impair the growth of these cells and slow down clonal expansion. In the light of this tradeoff, what is the overall effect of CIN on the establishment and progression of a cancer? This is the subject of the current chapter.

6.1 The effect of chromosome loss on the generation of cancer

We study the role of chromosomal instability in the context of the inactivation of tumor suppressor genes (TSP). We will concentrate on a specific event, namely, the chromosome loss event [Thiagalingam *et al.* (2001)]. Other features of CIN such as mitotic recombination or chromosome duplication, may contribute to an activation of oncogenes or gene dosage effects [Luo *et al.* (2000); Tischfield and Shao (2003); Wijnhoven *et al.* (2001)], but such events cannot turn off a TSP. Thus, focusing on cancers with a TSP allows us to isolate one feature of CIN, and identify its role in cancer progression.

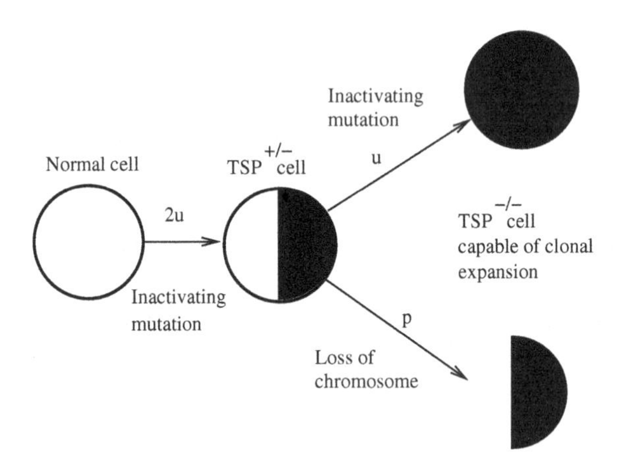

Fig. 6.1 Two mechanisms of a TSP inactivation. First, one allele of the TSP must be inactivated by a small-scale event, e.g. a point mutation. Then, there are two possibilities. Either the second allele experiences another small-scale hit (the phenotype with two inactivated copies of the gene is represented by a black circle). Or, the whole chromosome containing the second, functional copy of the TSP could be lost (this phenotype is represented by a black semicircle).

Let us start our quantitative study by identifying exactly how loss of chromosomes may influence the inactivation of a TSP, see Figure 6.1 [Komarova and Wodarz (2004)]. In a normal cell (an empty circle), both maternal and paternal chromosomes are present, and both alleles of the TSP are intact. An inactivating mutation can occur which turns off one of the alleles of the TSP (this is represented by a half of the circle turning black). The corresponding phenotype is denoted by $\text{TSP}^{+/-}$. For "classical" TSP's there is no noticeable change in function of such cells; both alleles must be inactivated before a phenotypic change is observed. This second event, the inactivation of the remaining allele of the TSP, can happen in two ways. First of all, another inactivating small-scale event could occur (both halves of the circle become black). Alternatively, the second allele can be lost by a loss-of-chromosome event (this is depicted by means of a "missing" half of a circle). This will unmask the mutated copy of the TSP and lead to a phenotypic change in the cell.

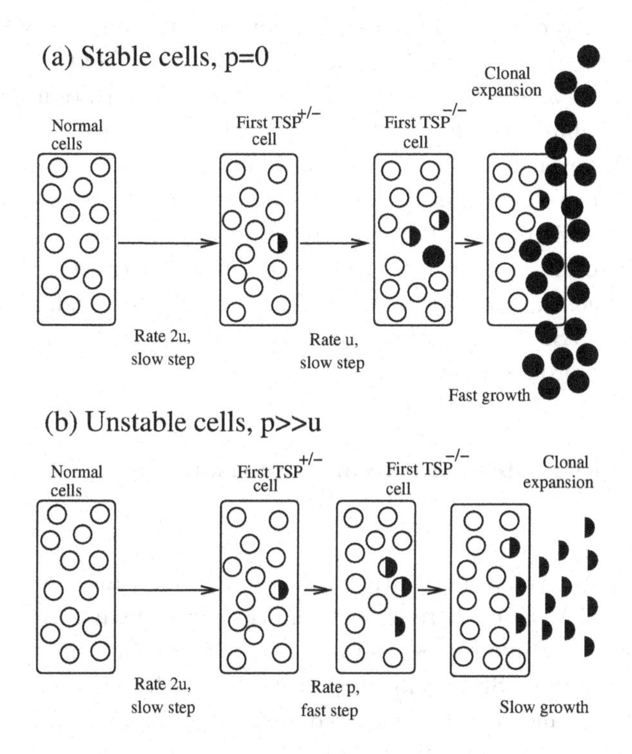

Fig. 6.2 TSP inactivation and clonal expansion. (a) In the case where chromosome losses occur rarely ($p = 0$), we have the following sequence of events: first, a mutant appears which has one copy of the TSP inactivated. After a while, a second mutation may occur producing a cell with two inactivated copies of the TSP. This leads to clonal expansion. (b) If losses of chromosomes are possible, then one of the progeny of the first $TSP^{+/-}$ cell may lose the chromosome containing the functional copy of the TSP, thus giving rise to cells with one inactivated TSP copy and one "missing" TSP copy. Such cells enter a phase of clonal expansion, but this happens at a slower rate compared to (a) because of frequent chromosome loss events resulting in dead or non-reproductive cells.

Let us denote the rate at which small-scale genetic events happen by u (per cell division per gene), and the rate of chromosome loss by p (per cell division per chromosome). The basic rate at which such mutation events occur in stable cells has been estimated to be approximately $u = 10^{-7}$ per cell division per gene. The inactivation of the first allele of the TSP will happen with the rate $2u$, because there are two alleles. The inactivation of the second allele can happen with the rate u by a mutation, and with the rate p by loss of chromosome, see Figure 6.2. Let us first suppose that the rate of chromosome loss is zero, $p = 0$, Figure 6.2a; a TSP gene can only be inactivated by two consecutive, independent (small scale) genetic events. This is possible, but the probability of such a double mutation is very low. Next, let us consider the opposite extreme, where the rate of LOH is very high, such that $p \gg u$, Figure 6.2b. Now, the second inactivation event happens with probability p, that is, it is greatly accelerated compared to the case $p = 0$. However, the price that the cell lineage has to pay is a very high rate at which non-viable mutants are produced. This will considerably slow down the expansion of the TSP-negative phenotype.

Therefore, there must be an intermediate, optimal (for cancer!) value of the rate of chromosome loss, for which wild-type cells have a high chance of inactivating the TSP gene, without having to pay too high a price in non-viable or non-reproductive mutants.

6.2 Calculating the optimal rate of chromosome loss

The model set-up. We model epithelial tissue organized into compartments. In the simplest case, there is one stem-cell per compartment. For example, in colon this would correspond to crypts with a stem-cell situated at the base of each crypt. Stem cells divide asymmetrically producing one (immortal) stem-cell and one differentiated cell. Here we concentrate on the dynamics of the stem cells. Each division event is equivalent to a replacement of the old stem cell with a copy of itself. Upon division of a stem cell, the immortal daughter cell might (i) acquire a silencing mutation in one of its alleles of the APC gene with probability u per cell division, or (ii) lose one of its chromosomes, with probability p per cell per cell division per chromosome. Once both copies of the TSP gene have been inactivated, the cell will be able to escape homeostatic control and create a growing clone. We will describe the clonal expansion by a deterministic model.

Uncertainties still exist about the exact cellular origins of cancer, see Chapter 5. According to the stem-cell theory, it is the stem cells which are at risk. The de-differentiation theory suggests that partially differentiated cells could be targets for cancerous mutations. If we do not want to restrict ourselves to one or the other theory, we can solve the problem of optimization for a number of different assumptions. According to one scenario, cancer is initiated in adult stem cells. There are or several stem cells per compartment, such as the crypt of the colon. Alternatively, we can assume that a healthy compartment contains a population of partially differentiated cells, which are subject to a constant turnover, but still maintain a constant size of the compartment. Depending on the number of cells, this can be described either with a stochastic or a deterministic model. Again, inactivation of a TSP gene results in clonal expansion. It turns out that the results remain very similar in the context of the different assumptions.

Optimal rate of chromosome loss. Suppose that a stem cell has a probability to lose a chromosome p per chromosome per cell division. First we calculate the probability to inactivate the TSP gene by time t. The sequence of events can be expressed by the following simple diagram,

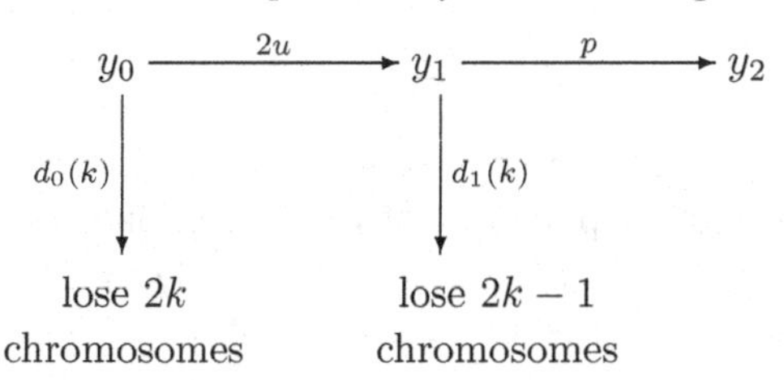

Here y_i is the probability for the stem cell to have i inactivated copies of the TSP gene. The first event of inactivation happens by a fine-scale genetic event (probability u times two for two alleles), and the second event is a loss of the chromosome with the remaining copy of the TSP gene (probability p). The parameter k is related to the cost of chromosome loss, as explained below.

A very important issue here is the exact cost of LOH events for the cell and its reproductive potential. In the most optimistic (for cancer) scenario, (a), there is no reduction in fitness due to the loss of any other chromosomes: the only chromosome that "counts" is the one containing the TSP gene. At stage y_0, a loss of either copy damages the cell, and at stage y_1, a loss of the chromosome with the *mutated* copy of the TSP gene is harmful (and a loss of the other copy leads to a clonal expansion). An alternative interpretation of this extreme case is that while loss of a single chromosome copy would reduce fitness, this is buffered by duplication events. In the most pessimistic scenario (b), a loss of any chromosome results in cell death, unless it leads to a TSP inactivation. It is safe to say that the reality is somewhere between these extreme scenarios.

For scenario (b), we set $d_0(k) = 1-(1-p)^{2k}$ and $d_1(k) = 1-(1-p)^{2k-1}$, where $k = 23$ is the number of chromosomes. For scenario (a), the death rates can be expressed by the same formulas with $k = 1$.

We can write down the Kolmogorov forward equations for all the probabilities (skipping the argument k of d_0 and d_1),

$$\dot{y}_0 = [(1-d_1)(1-2u) - 1]\, y_0, \tag{6.1}$$

$$\dot{y}_1 = (1-d_0)2uy_0 + [(1-d_1)(1-p) - 1]\, y_1, \tag{6.2}$$

$$\dot{y}_2 = (1-d_1)py_1, \tag{6.3}$$

with the initial condition $y_0(0) = 1$. We need to calculate the probability distribution of creating a TSP$^{-/-}$ mutant as a function of time, which is given by $\dot{y}_2$. We have,

$$\dot{y}_2(t) = \frac{up(e^{-ut} - e^{-(p+d_1)t})}{p + d_1 - u}, \tag{6.4}$$

where we assumed that $ut \ll 1$. Note that the argument given here holds without change for (constant) populations of more than one cell, as long as the number N of cells satisfies $N < 1/u$ and $N < 1/\sqrt{p}$. Otherwise, the calculations can be easily adapted to include the effect of tunneling, see Chapter 3.

Once a TSP$^{-/-}$ cell has been produced, it starts dividing according to some law which is (at least, initially) close to exponential. Starting from one cell at time $t = 0$, by time t we will have $Z_y(t)$ cells, with

$$Z_y(t) = e^{a\beta[1-d_1(k)]t}. \tag{6.5}$$

The parameter a is the growth rate of the initiated cells, and $0 < \beta < 1$ is the cost due to the fact that a chromosome is missing from all CIN cells because of the inactivation of the TSP by a loss of chromosome. The factor $[1 - d_1(k)]$ comes from the probability for a CIN cell to produce a non-viable mutant, which for scenario (a) only happens if only one particular chromosome is lost, and for scenario (b) - if any chromosome is lost.

If we now include the mutation stage, we will need to evaluate the convolution,

$$Z_y(t) = \int_0^t \dot{y}_2(t')e^{a\beta(1-p)^{2k-1}(t-t')}\, dt'.$$

The integral yields the following law of growth:

$$Z_y(t) = \frac{upe^{a\beta(1-d_1)t}}{a\beta(1-d_1)},$$

where we assumed that for relevant times, $a\beta t > 1$.

Let us ask the following question: *how long does it take, on average, for a TSP$^{-/-}$ clone to reach a certain size?* The answer will depend on all the parameters of the system, and in particular, on the rate of chromosome loss, p. For the reasons explained above, the waiting time will be very large both for $p = 0$ and for very high values of p. Indeed, for very small p the mutations that lead to a TSP inactivation will take too long, and for very large values of p, the clonal expansion will be too slow because of the amount of non-viable or non-reproductive cells produced. The waiting time will have a minimum for an intermediate value of $p = p_*$, which we call the *optimal* (for cancer) value of the rate of chromosome loss. With this value of p, a cancer will appear and grow at the fastest rate. This approach is equivalent to the "minimum-time-to-target" method in optimization theory.

To find an optimal value of p that maximizes the growth, we solve $Z_y(t) = M$ for t, which gives,

$$t(M) = \frac{1}{a\beta(1-d_1)} \log\left[\frac{a\beta M}{u}\frac{1-d_1}{p}\right],$$

and then we minimize this as a function of p. This can be done easily if we assume that $p \ll 1/(2k)$ (it will turn out that the result for p_* satisfies this assumption). Expanding the expression $dt(M)/dp$ in terms of p, we obtain the equation for p,

$$\frac{1}{p} = (2k-1)\log\frac{a\beta M}{up},$$

where we formally have $k = 1$ for scenario (a), and $k = 23$ for scenario (b).

Large initial number of cells. In the above model, the number of wild-type cells in the compartment is small ($N \ll 1/u$). In order to handle the scenario where a large number of cells are competing in a compartment, which may correspond to later stages of carcinogenesis, we numerically simulated a set of quasispecies–type equations. The estimate obtained for the optimal value of p is very similar to the ones given for the stochastic model above.

Parameter dependence of the result. The result for p_* turns out to be amazingly robust, see Table 6.1. We can see that p_* depends logarithmically (that is, weakly), on the combination $\kappa = a\beta M/u$. As we vary these parameters over many decades, so that κ changes from 10^5 to 10^{20}, the result for the optimal value of p varies only slightly. Interestingly, it also does not significantly depend on the overall fitness cost for the cell brought about by chromosome loss. The results for scenarios (a) and (b) are presented in Table 6.1. The optimal value of p is lower in scenario (b), which is not surprising because this case assumes a higher penalty for chromosome loss events. The remarkable fact is that the values of p_* for the two scenarios are so close to each other, and that they depend so little on the assumptions of the model.

Table 6.1 Calculated values for the optimal rate of chromosome loss, p_*, for different values of the parameter, κ, and for each of the two scenarios, (a) and (b).

Scenario	$\kappa = 10^{20}$	$\kappa = 10^5$
(a) Optimistic	2×10^{-2}	8×10^{-2}
(b) Pessimistic	5×10^{-4}	2×10^{-3}

What is even more encouraging is that we can compare these results with the value of the rate of chromosomal loss obtained by Lengauer et al. [Lengauer *et al.* (1997)] in vitro for several CIN colon cancer cell lines. In their paper, Lengauer et al. allowed cell colonies to grow from a single cell for 25 generations, after which FISH analysis was performed on a subset of the progeny. This allowed to count the number of individual chromosomes in cells. The average number of chromosomal copies was calculated for each cell line, for each chromosome, and this was compared with the mode number, equivalent to the number of chromosome copies in the original cell. This was the first (and only) experiment which allowed to calculate the *rate* of chromosome loss and gain, as opposed to the estimates of the *frequency* of various chromosomal aberrations in a given lesion/cell colony. Two types of cancer cells have been used: some known to possess mismatch repair instability, and some characterized by CIN. In the cell lines with microsatellite instability, the rate of chromosome loss was the same as control (and indistinguishable from the background). In the chromosomally unstable cell lines, the rate of chromosome copy change was highly elevated. The value that emerges from experiments of Lengauer et al. is $p = 10^{-2}$ per chromosome per cell division, which is almost exactly in the middle of the range that we obtained theoretically.

6.3 Why does CIN emerge?

Next, we will discuss the ways by which CIN could come about in carcinogenesis. Let us compare two cell lines, one with CIN, such that its rate of chromosome loss is optimal, $p = p_*$, and another without CIN, such that $p \ll p_*$. From our argument above, and from the definition of the optimal rate, p_*, it is clear that the unstable cell line will grow faster. Now let us reformulate the question slightly. Suppose that we start from a non-CIN wild-type cell. In order to use the "advantages" of CIN, a cell must at some points acquire the CIN phenotype.

Comparison of stable and unstable pathways. Let us include the step of initiation of CIN. All pathways can be expressed by the following diagram,

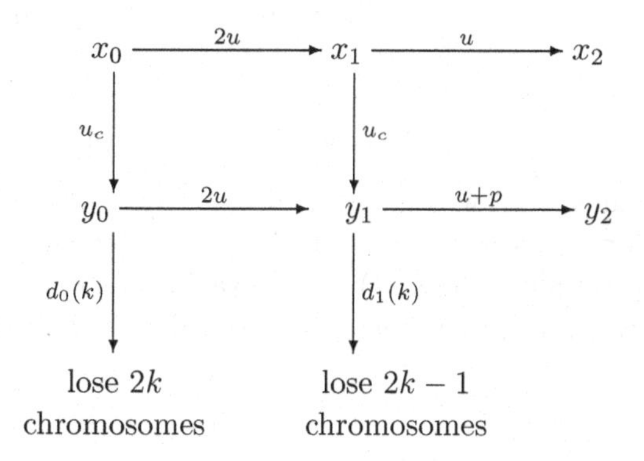

Here x_i are the probabilities for the stem cell to be stable and have i inactivated copies of the TSP gene, and y_i are the probabilities for the cell to be CIN and have i inactivated copies of the TSP gene. u_c is the rate at

which a cell acquires CIN. The Kolmogorov forward equations are:

$$\dot{x}_0 = [(1-2u)(1-u_c)-1]x_0, \tag{6.6}$$
$$\dot{x}_1 = (1-u_c)2ux_0 + [(1-u_c)(1-u)-1]x_1, \tag{6.7}$$
$$\dot{x}_2 = (1-u_c)ux_1, \tag{6.8}$$
$$\dot{y}_0 = (1-u)u_c x_0 + [(1-d_1)(1-2u)-1]\,y_0, \tag{6.9}$$
$$\dot{y}_1 = (1-d_0)2uy_0 + (1-u)u_c x_1 + [(1-d_1)(1-u-p)-1]\,y_1, \tag{6.10}$$
$$\dot{y}_2 = (1-d_1)(u+p)y_1, \tag{6.11}$$

with the initial condition $x_0(0) = 1$. It is easy to show that for the optimistic (for cancer) scenario (a), the two CIN pathways ($x_0 \to y_0 \to y_1 \to y_2$ and $x_0 \to x_1 \to y_1 \to y_2$) contribute equally to y_2. For the pessimistic (for cancer) scenario (b), and $p \gg u$, the second of these pathways gives a much larger contribution. The reason for this is that losing a "wrong" chromosome will destroy the cell line in this extreme scenario. Therefore, it is much more likely to reach the state y_2 if CIN appears as late as possible. In what follows we will ignore the first pathway entirely because it either does not contribute anything or gives a factor of 2. This simplifies the calculation because now, the first step for both stable and CIN cancer is $x_0 \to x_1$, and if we only want to compare the CIN and non-CIN pathways with each other, this step can be ignored. This is equivalent to starting from $x_1(0) = 1$ rather than $x_0(0) = 1$.

The probability distribution of creating a TSP$^{-/-}$ mutant as a function of time, is given by $\dot{x}_2$ for the stable pathway and by $\dot{y}_2$, for the unstable pathway. We have,

$$\dot{x}_2(t) = ue^{-u_c t},$$

and $\dot{y}_2$ is given by formula (6.4), with u replaced by u_c.

The clonal expansion law for unstable cells is given by equation (6.5), and for stable cells we simply have $Z_x(t) = e^{at}$. Taking a convolution of the rates for the mutation and expansion stages, we arrive at the following laws of growth:

$$Z_x(t) = \frac{ue^{at}}{a}, \quad Z_y(t) = \frac{u_c(u+p)e^{a\beta(1-d_1)t}}{a\beta(1-d_1)}.$$

It turns out that unless u_c is several orders of magnitude bigger than u, Z_y grows slower than Z_x, that is, genetically activated CIN cannot be advantageous.

Can CIN be the first event? Here is what the calculation above shows. Let us assume that

(1) CIN is a genetic event which happens at a rate comparable to the basic mutation rate, u, or even a couple of orders of magnitude larger, and
(2) unstable cells do not have any additional fitness advantages compared to wild type cells.

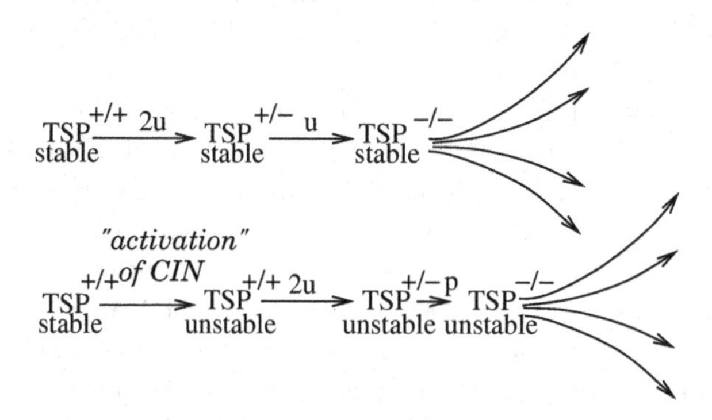

Fig. 6.3 Cancer initiation and progression in the case of stable cells (above), and unstable cells (below). The latter pathway includes an additional step, "activation of CIN". Also, the inactivation of the second copy of the TSP gene is a fast step in the case of unstable cancers. The second pathways takes longer if conditions (a) and (b) are satisfied (see text).

Then a CIN cancer, even with the optimal value of p_*, still cannot progress faster than a stable cancer, simply because it requires this extra event, the "activation" of the CIN phenotype. Let us outline the basic reasoning at an intuitive level, by comparing two sequences of events, see Figure 6.3. The first one (for stable cancers) involves two slow inactivation events and clonal expansion. The other (for unstable cancers) involves some mutations leading to the acquisition of the unstable phenotype, two inactivation events (one slow and the other fast), and clonal expansion. Our calculations show that under assumption (1) and (2) above, the second sequence of events can never happen faster than the first one.

The following conclusions can be drawn from this argument. In order for CIN to be selected for, that is, to play the role of the driving force in cancer progression, at least one of the assumptions, (1) or (2), must be wrong. Let us first explore the possibility that (1) is violated, and then move on to (2).

One way to see how CIN could accelerate cancer progression is to assume that CIN comes about by some epigenetic mechanism with a rate much faster than the basic mutation rate [Eden *et al.* (2003); Gaudet *et al.* (2003); Lindblom (2001)]. This hypothesis is consistent with the numerous but still unsuccessful attempts to find a "CIN gene" [Amon (1999); Gemma *et al.* (2000); Kolodner *et al.* (2002); Michel *et al.* (2001); Ohshima *et al.* (2000); Wassmann and Benezra (2001)] and (at least, partially) epigenetic nature of CIN in yeast. If this were true, then the "activation of CIN" step in the diagram of Figure 6.3 would be short, and this would give CIN a chance to be "beneficial" for cancer.

Another possibility is that CIN arises because of alternative reasons, such as environmental factors, so that the unstable phenotype has an advantage compared to the wild type cells. For example, if cells are exposed to high degrees of DNA damage (as a result of carcinogens and metabolic radicals), CIN can be selected for, because it avoids frequent cell cycle arrest upon damage [Gasche *et al.* (2001)]. The effect of DNA damage on the selection of genetically unstable cells is the subject of the next chapter. In this case, all the steps in the diagram for unstable cancer (Figure 6.3) will be accelerated, which means that the instability indeed facilitates progression to cancer.

Alternatively, CIN might be the consequence of another mutation which confers an advantage to the cell [Cahill *et al.* (1999)]. It has been proposed that a mutation in the APC gene itself leads to the development of CIN and the generation of aneuploidy in colon cancer [Fodde *et al.* (2001a); Kaplan *et al.* (2001)]. This could lead to a number of possibilities, for instance, two steps in the pathway for the unstable cancer combined in one.

6.4 The bigger picture

We have calculated the optimal rate of chromosome loss assuming that cancer is initiated by the inactivation of a TSP followed by a clonal expansion. The resulting rate, $p_* \approx 10^{-2}$ per cell division per chromosome, is similar to that obtained experimentally by Lengauer et al. This is a thought-provoking result. A hypothesis consistent with our finding is that the rate at which cancerous cells lose chromosomal material is under selection pressure, and as a result, the optimal rate, p_*, is the one that survives the competition. In other words, out of many possibilities, we will mostly see the cancers that have the optimal rate of chromosome loss, because these are cancers that are initiated and progress at the fastest rate.

The next natural question is the following. What happens if there are more TSP's which need to be inactivated down the line, as cancer progresses? It is easy to see that adding another TSP to the pathway will not change the value of p_* significantly. This is because every new TSP gene takes much less time to be inactivated than a previous one; this is a consequence of a growing size of the lesion.

The optimal rate of chromosome loss calculated in this chapter is indeed optimal during early and intermediate stages of cancer progression as long as they involve TSP's. However, at later stage of carcinogenesis, the selective pressures optimizing the rate of LOH change drastically. It is well-known that a lesion cannot grow above a certain small size (about $2\,mm$) without extra blood supply (angiogenesis). A larger or metastasizing tumor is hard to maintain, and the price of losing chromosomes becomes too high to be balanced by an elevated variability. Therefore, we predict that at later stages, the optimal rate of LOH will decrease to nearly zero [Komarova (2004)]. This is consistent with observations that late stage cancers are sometimes (surprisingly) stable. In their recent review, Albertson et al. [Albertson *et al.* (2003)] note that chromosome aberration spectra seem to stabilize in advanced cancers. Some evidence comes from comparing tumor genomes (in the same individual) of in situ and invasive lesions [Kuukasjarvi *et al.* (1997b)], primary and recurrent tumors [Waldman *et al.* (2000)] and primary and metastatic tumors [Kuukasjarvi *et al.* (1997a)]. Also, some established cancerous cell lines exhibit remarkable stability [Yoon *et al.* (2002)], suggesting that they may originate from a system where an optimal, stable phenotype has been shaped by selective forces.

Similarly, if the activation of oncogenes (rather than the inactivation of tumor suppressor genes) plays a major role in the progression of cancers, then chromosomal instability is likely to be detrimental to the cancer. Indeed, to turn on an oncogene, a small scale mutation is often needed rather than a chromosome loss event or another crude chromosomal change. Moreover, a chromosome loss event may lead to the inactivation of a functioning oncogene which will revert the process of oncogenesis. Further mathematical work will be required to investigate the effect of CIN in the context of both oncogenes and tumor suppressor genes.

To leave the reader with an important message, we note that our analysis leads to the insight that CIN does not arise simply because it allows a faster accumulation of carcinogenic mutations. Instead, CIN must arise because of alternative reasons, such as environmental factors, fast/epigenetic events, or as a direct consequence of a TSP inactivation. The increased variability alone is not a sufficient explanation for the presence of CIN in the majority of cancers.

Chapter 7

DNA damage and genetic instability

Cancer is initiated and progresses via the accumulation of multiple mutations. The last chapters presented mathematical analyses of how cells proceed down this pathway to cancer in the most efficient way. In particular, the question was addressed whether genetic instability is observed because it allows cells to acquire oncogenic mutations at a faster rate. The mathematical approaches concentrated entirely on the cells which develop cancer, and did not take into account environmental factors. Environmental factors can greatly influence whether cells can become cancerous and grow successfully. They may also provide conditions under which genetically unstable cells have a selective advantage. Identifying such conditions is important, since they might contribute to explaining why so many cancers are characterized by genomic instability.

A major environmental factor in the development of cancer is the amount of DNA damage which cells experience. DNA damage can come from a variety of sources. Carcinogens contained in food or in the air we breath can damage DNA. UV radiation can break DNA. Chemotherapeutic agents can lead to various forms of DNA damage. Most importantly perhaps, aging leads to an increased amount of DNA damage. This is because metabolic activities produce reactive oxygen species which are toxic for our genome [Campisi (2001)].

How does DNA damage influence the process of carcinogenesis? On the most elementary level, it might increase the basic mutation rate. More damage results in a higher probability that mutations are produced. More profoundly, however, it might also influence whether genetically unstable cells are more advantageous than stable cells, or whether stable cells can grow better than unstable ones. High amounts of DNA damage have the following consequences for **stable cells**. On the one hand, the cells maintain stable genomes. On the other hand, each time a cell gets hit it repairs the damage. This takes time and is usually manifested in cell cycle arrest or in stalling of the replication process. Repair is therefore costly because it slows down the overall growth of the cell population. **Unstable cells** are influenced by high levels of DNA damage in the following way. They avoid repair and therefore do not enter cell cycle arrest. On the other hand, they pay an alternative cost. Many mutants are created, and a large proportion of the mutants are likely to be non-viable.

This chapter will present a mathematical model to investigate whether and how DNA damage can influence the growth processes of stable and unstable cells. This is done by examining the competition dynamics between stable and unstable cells. Which cell type wins? Can an increase in the level of DNA damage reverse the outcome of competition?

7.1 Competition dynamics

We start by exploring the competition dynamics between a stable and a mutator cell population [Komarova and Wodarz (2003)]. They differ in the probability with which they repair genetic damage. Stable cells repair damage with a probability ϵ_s, and mutator cells repair damage with a probability ϵ_m, where $\epsilon_s > \epsilon_m$. We further assume that these cell populations differ in their intrinsic rate of replication. The stable cells replicate at a rate r_s, and the mutator cells replicate at a rate r_m. Let us denote the abundance of stable and mutator cells as S and M, respectively. The competition dynamics are given by the following pair of differential equations which describe the development of the cell populations over time,

$$\dot{S} = r_s S(1 - u + \beta\epsilon_s u) + \alpha u r_s S(1 - \epsilon_s) - \phi S, \tag{7.1}$$

$$\dot{M} = r_m M(1 - u + \beta\epsilon_m u) + \alpha u r_m M(1 - \epsilon_m) - \phi M. \tag{7.2}$$

The model is explained graphically in Figure 7.1. The cells replicate at a rate r_s or r_m. These parameters reflect how often cells reproduce and die; we will call this the *intrinsic replication rate* of the cells. The two cell populations compete for a shared resource. Competition is captured in the expressions ϕS and ϕM, where ϕ is defined as follows:

$$\phi = Sr_s\left[1 - u\Big(1 - \beta\epsilon_s - \alpha(1 - \epsilon_s)\Big)\right] + Mr_m\left[1 - u\Big(1 - \beta\epsilon_m - \alpha(1 - \epsilon_m)\Big)\right].$$

During replication a genetic alteration can occur with a probability u.

(a)

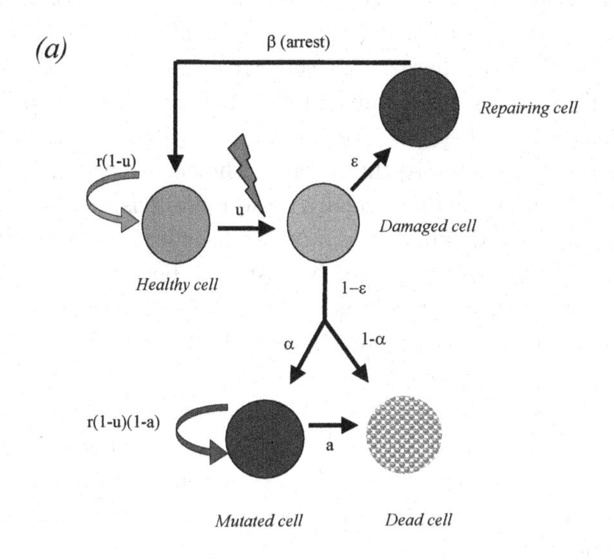

(b)

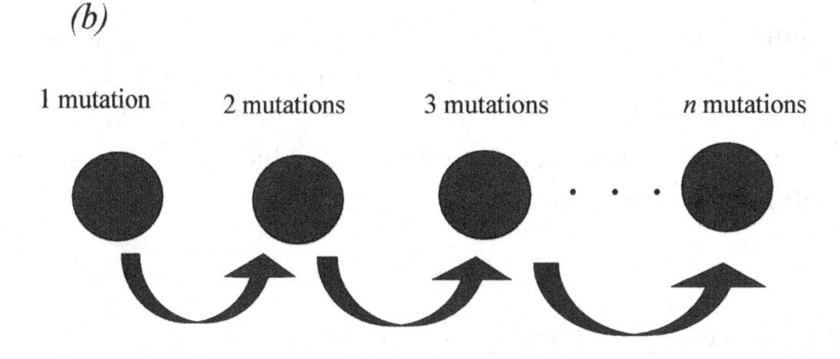

Fig. 7.1 Schematic Diagram of the model. (a) Process of cell reproduction, DNA damage, repair, cell cycle arrest, mutation, and death. (b) When DNA damage is not repaired, the cells can accumulate mutations. In the model cancer progression corresponds to the successive accumulation of mutations, also referred to as the mutation cascade.

We call this the DNA hit rate. DNA damage can occur both spontaneously (most likely at low levels), or it can be induced by DNA damaging agents which corresponds to a high value of u. If a genetic alteration has occurred, it gets repaired with a probability ϵ_s or ϵ_m. During repair, there is cell cycle arrest, and this is captured in the parameter β. The value of β can lie between zero and one and thus reduces the rate of cell division (given by βr). If $\beta = 0$, the repairing cells never replicate and this is the maximal cost. If $\beta = 1$, there is no cell cycle arrest and no cost associated with repair. With a probability ($1 - \epsilon_s$ or $1 - \epsilon_m$) the genetic alteration does not get repaired. If the alteration is not repaired, a mutant is generated. A mutation is therefore the result of the occurrence of DNA damage combined with the absence of repair. The mutant is viable (and neutral) with a probability α, while it is non-viable with a probability $1 - \alpha$. Therefore, the model captures both the costs and benefits of repair: Efficient repair avoids deleterious mutations but is associated with cell cycle arrest. Absence of efficient repair can result in the generation of deleterious mutants, but avoids cell cycle arrest.

Note that in this first model, we assume that mutants that are created are either non-viable (and thus do not participate in the competition dynamics) or neutral (and thus have the same intrinsic reproductive rate as the wild type). We will include the possibility of advantageous and disadvantageous (but viable) mutants later.

Le us explore how the competition dynamics depends on the rate at which cells acquire genetic alterations (DNA hit rate, u). In general, if two cell populations compete, the cells with the higher fitness wins. The fitness of the cells is given by $r_{s,m}[1 - u[1 - \alpha + \epsilon_{s,m}(\alpha - \beta)]]$. Note that the quantity $1 - \alpha$ has the meaning of the cost of production of deleterious mutants; we will refer to it as

$$C_{del} = 1 - \alpha.$$

Similarly, the quantity $1 - \beta$ is the cost of cell cycle arrest,

$$C_{arr} = 1 - \beta.$$

In these notations, we can rewrite the expression for the fitness in a more intuitive way,

$$r_{s,m} - ur_{s,m}[C_{del} + \epsilon_{s,m}(C_{arr} - C_{del})]. \tag{7.3}$$

If the DNA hit rate is low (low value of u), the fitness of the cells is approximately given by their intrinsic rate of replication (r_s and r_m). Thus, the cell population with the higher intrinsic replication rate has a higher fitness than the cell population with the lower intrinsic replication rate. On the other hand, when the DNA hit rate, u, is increased, the fitness

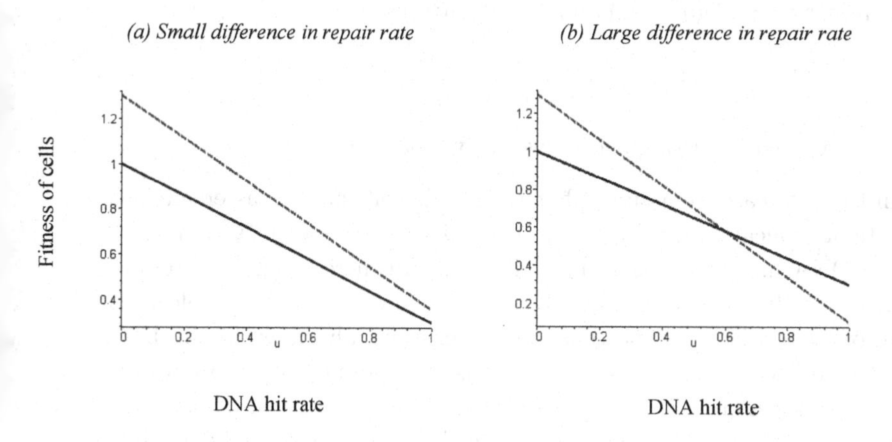

Fig. 7.2 Effect of the DNA hit rate, u, on the fitness of two cell populations. At low DNA hit rates, the population with the higher intrinsic replication rate wins. An increase in the DNA hit rate decreases the fitness of both cell populations. However, the degree of fitness reduction of the population characterized by the higher intrinsic replication rate is stronger than that of the slower population of cells. If there is a sufficient difference in the repair rates (degrees of genetic stability) between the two cell populations (a), an increase in the DNA hit rate can result in a reversal of the relative fitnesses, and thus in a reversal of the outcome of competition. If the difference in repair rates between the two cell populations is not sufficient, (b), we do not observe such a reversal. Parameter values were chosen as follows: $r_s = 1$; $r_m = 1.3$; $\alpha = 0.05$; $\beta = 0.3$; $\epsilon_s = 0.99$. For (a) $\epsilon_m = 0.1$. For (b) $\epsilon_m = 0.9$.

depends more strongly on other parameters. In particular, the fitness of both populations can depend on the DNA hit rate, u. Notably, an increase in the value of u may result in a stronger decline in fitness of the cell population with the faster intrinsic rate of replication relative to the slower cell population (Figure 7.2). Therefore, if the DNA hit rate crosses a critical threshold, $u > u_c$, the outcome of competition can be reversed. The value of u_c is given by

$$u_c = \frac{r_s - r_m}{(r_s - r_m)C_{del} + (r_s\epsilon_s - r_m\epsilon_m)(C_{arr} - C_{del})}. \tag{7.4}$$

We are interested to find out, under what circumstances reversal of competition can occur. One condition required for the reversal of competition is that the stable and mutator cells are characterized by a sufficient difference in the repair rate (Figure 7.2) which is defined as

$$\Delta\epsilon \geq \frac{|r_s - r_m|[(1 - C_{del})(1 - \epsilon_s) + \epsilon_s(1 - C_{arr})]}{|C_{arr} - C_{del}|r_m}.$$

Further, we need to distinguish between two scenarios.

(1) In the first case we assume that the stable cells have a faster intrinsic rate of replication than the mutator cells (i.e. $r_s > r_m$). Therefore, at low DNA hit rates, the stable cells win. An increased DNA hit rate, u, can shift the competition dynamics in favor of the unstable cells. In other words, unstable cells gain a selective advantage as the DNA hit rate becomes large. This is because the population of stable cells frequently enters cell cycle arrest when repairing genetic damage and this slows down the overall growth rate. For this outcome to be possible, the following condition has to be fulfilled: The cost of cell cycle arrest, C_{arr}, must be greater than the cost of producing non-viable mutants, C_{del}. If this condition is not fulfilled, reversal of competition at high DNA hit rates is not observed.

(2) In the second case we assume that the stable cells have a slower intrinsic replication rate than the mutator cells (i.e. $r_s < r_m$). Therefore, at low DNA hit rates, the unstable cells win. An increased DNA hit rate, u, can shift the competition dynamics in favor of the stable cells. In other words, a high DNA hit rate selects against genetic instability. This is because the unstable cells produce more non-viable mutants and this reduces the effective growth rate significantly. In contrast to the previous scenario, this requires that the cost of producing non-viable mutants, C_{del}, must be higher than the cost of cell cycle arrest, C_{arr}. If this condition is not fulfilled, reversal of competition at high DNA hit rates is not observed.

Table 7.1 Summary of the basic competition dynamics. If the mutators (M) have a lower intrinsic replication rate than the stable cells (S), a high DNA hit rate can select in favor of M. If the intrinsic replication rate of M is higher than that of S, then a high DNA hit rate can select for S.

	M slower than S	**M faster than S**
Low DNA hit rate	S win	M win
High DNA hit rate	M win if $C_{arr} > C_{del}$	S win if $C_{arr} < C_{del}$

To summarize, this analysis gives rise to the following results (Table 7.1). A high DNA hit rate, u, can reverse the outcome of competition in favor of the cell population characterized by a slower intrinsic growth rate if the competing populations are characterized by a sufficient difference in their repair rates. The higher the difference in the intrinsic replication rate of the two cell populations, the higher the difference in repair rates required to reverse the outcome of competition. If the intrinsic replication rate of the genetically unstable cell is slower, a high DNA hit rate can select in favor of genetic instability. On the other hand, if the intrinsic growth rate of the genetically unstable cell is faster, a high DNA hit rate can select against genetic instability.

7.2 Competition dynamics and cancer evolution

7.2.1 *A quasispecies model*

In the previous section, we considered the competition dynamics between stable and unstable populations of cells, assuming that they are characterized by different and fixed rates of replication. We further assumed that mutations are either non-viable or neutral. However, mutations are unlikely to be neutral, and will change the replication rate of the cells. In other words, cells may evolve to grow either at a faster or a slower rate, depending on the mutations generated. Here, we extend the above model to take into account such evolutionary dynamics.

The competition problem. As before, we consider two competing cell populations: a genetically stable population, S, and a mutator population, M, see Figure 7.1b. We start with unaltered cells which have not accumulated any mutations. They are denoted by S_0 and M_0, respectively. Both are assumed to replicate at the same rate, r_0. When the cells become damaged and this damage is not repaired, mutants are generated. If the mutants are viable, they can continue to replicate. During these replication events, further mutations can be accumulated if genetic alterations are not repaired. We call the process of accumulation of mutations the *mutational cascade*. Cells which have accumulated i mutations are denoted by S_i and M_i, respectively, where $i = 1, \ldots, n$. They are assumed to replicate at a rate r_i. Stable and unstable cells differ in the rate at which they proceed down the mutational cascade. In addition to the basic dynamics of cell replication described in the previous section, we assume that during cell division, mutated cells can undergo apoptosis, since oncogenic mutations can induce apoptotic checkpoints [Seoane *et al.* (2002); Vogelstein *et al.*

(2000b)]. Thus, the intrinsic replication rate of mutated cells is given by $r_i(1-a)$, where a denotes the probability to undergo apoptosis upon cell division. These processes can be summarized in the following equations:

$$\dot{S}_0 = R_0 S_0(1-u_s) - \phi S_0, \tag{7.5}$$

$$\dot{S}_i = \alpha u R_{i-1} S_{i-1}(1-\epsilon_s) + R_i S_i(1-u_s) - \phi S_i, \quad 1 \leq i \leq n-1, \tag{7.6}$$

$$\dot{S}_n = \alpha u R_{n-1} S_{n-1}(1-\epsilon_s) + R_n S_n[1-u_s+\alpha u(1-\epsilon_s)] - \phi S_n, \tag{7.7}$$

$$\dot{M}_0 = R_0 M_0(1-u_m) - \phi M_0, \tag{7.8}$$

$$\dot{M}_i = \alpha u R_{i-1} M_{i-1}(1-\epsilon_m) + R_i M_i(1-u_m) - \phi M_i, \quad 2 \leq i \leq n-1, \tag{7.9}$$

$$\dot{M}_n = \alpha u R_{n-1} M_{n-1}(1-\epsilon_m) + R_n M_n[1-u_m+\alpha u(1-\epsilon_m)] - \phi S_n, \tag{7.10}$$

$$\dot{w} = (1-\alpha)u\left[(1-\epsilon_s)\sum_{i=1}^{n} R_i S_i + (1-\epsilon_m)\sum_{i=1}^{n} R_i M_i\right] - \phi w, \tag{7.11}$$

where we introduced the following short hand notations: R_i is the effective intrinsic reproductive rate, $R_i = r_i(1-a)$ for $1 \leq i \leq n$ and $R_0 = r_0$, and $u_{s,m}$ are the two effective mutation rates, $u_{s,m} = u(1-\beta\epsilon_{s,m})$. The variable w denotes the non-viable mutants produced by the cells. The equations are coupled through the function ϕ, the average fitness, which is given by

$$\phi = (1-u_s)\sum_{j=0}^{n} R_j S_j + (1-u_m)\sum_{j=1}^{n} R_j M_j.$$

Solving quasispecies equations. Equations (7.5-7.11) are an example of a *quasispecies*-type system, which is a well-known population dynamical model in evolutionary biology. Quasispecies equations were first derived for molecular evolution by M. Eigen and P. Schuster [Eigen and Schuster (1979)], and since then have found applications in many areas of research, including biochemistry, evolution, and game theory.

In order to analyze system (7.5-7.11), we would like to review some of the techniques for solving quasispecies equations. Let the variable $\mathbf{x} = (x_0, x_1, \ldots, x_{n+1})$ satisfy the system

$$\dot{x}_0 = a_0 x_0 - \phi x_0, \tag{7.12}$$

$$\dot{x}_i = b_i x_{i-1} + a_i x_i - \phi x_i, \qquad 1 \leq i \leq n, \tag{7.13}$$

$$\dot{x}_{n+1} = \sum_{i=0}^{n} c_i x_i - \phi x_{n+1}, \tag{7.14}$$

where

$$\phi = (a_0 + c_0)x_0 + \sum_{k=1}^{n}[(a_k + c_k)x_k + b_k x_{k-1}].$$

We have $\sum_{k=0}^{n+1} x_k = 1$. System (7.12-7.14) is nonlinear. However, the nonlinearity can be removed by the following trick. Let us consider the variable $\mathbf{z} = (z_0, z_1, \ldots, z_{n+1})$ which satisfies the following system:

$$\dot{z}_0 = a_0 z_0, \tag{7.15}$$

$$\dot{z}_i = b_i z_{i-1} + a_i z_i, \qquad 1 \le i \le n, \tag{7.16}$$

$$\dot{z}_{n+1} = \sum_{i=0}^{n} c_i z_i. \tag{7.17}$$

If we set

$$x_i = z_i / \sum_{k=0}^{n+1} z_k, \quad 0 \le i \le n+1, \tag{7.18}$$

then the variable $\mathbf{x}$ satisfies system (7.12-7.14). The general solution of system (7.15-7.17) is given by

$$\mathbf{z}(t) = \sum_{j=0}^{n} \alpha_j \mathbf{v}^{(j)} e^{a_j t} + \alpha_{n+1}\mathbf{v}^{(n+1)}, \tag{7.19}$$

where α_j are constants determined from the initial condition, and $\mathbf{v}^{(j)}$ are eigenvectors of the appropriate triangular matrix corresponding to the eigenvalues a_j. The eigenvector $\mathbf{v}^{(n+1)} = (0, 0, \ldots, 0, 1)^T$ corresponds to the zero eigenvalue, and for the rest of the eigenvectors we have,

$$v_i^{(j)} = \begin{cases} 0, & i < j, \\ 1, & i = j, \\ (-1)^{i-j}\prod_{k=1}^{i-j}\frac{b_{j+k}}{(a_{j+k}-a_j)}, & j+1 \le i \le n. \end{cases} \tag{7.20}$$

From solution (7.19) and transformation (7.18) we can see that as time goes to infinity, the solution $\mathbf{x}(t)$ tends to the normalized eigenvector corresponding to the largest of the eigenvalues $a_0, \ldots, a_n$.

The exact solution corresponding to the initial condition $\mathbf{z}(0) = (1, 0, \ldots, 0)^T$ can be found. The appropriate coefficients in equation (7.19) are

$$\alpha_j = \prod_{m=1}^{j} \frac{b_m}{(a_j - a_{m-1})}, \quad 0 \leq i \leq n.$$

For the ith component, we obtain,

$$z_i(t) = \left(\prod_{k=1}^{i} b_k\right) \sum_{j=0}^{i} e^{a_j t} \prod_{m=0,\, m\neq j}^{i} \frac{1}{a_j - a_m}, \quad 1 \leq i \leq n. \tag{7.21}$$

The short-time behavior of this quantity is given by

$$z_i(t) = \left(\prod_{k=1}^{i} b_k\right) t^i, \quad 1 \leq i \leq n, \quad a_j t \ll 1 \quad \forall j.$$

The expression for z_{n+1} can also be obtained but is slightly more cumbersome.

Mutation cascades. Let us assume that $a_0 < a_n$, and in addition we have $a_i - a_{i-1} \sim a_i \sim 1$. Then the system exhibits the following behavior (Figure 7.3).

Starting from the "all x_0" state, the fraction of x_0 goes down steadily, and the population acquires some amount of x_1 (they may or may not be the majority). Upon reaching a maximum, the fraction of x_1 decreases and the fraction of x_2 experiences a "hump", to be in turn replaced by x_3, etc. The characteristic time at which each type experiences its maximum abundance can be estimated if we replace the expressions for $z_j(t)$ in (7.21) by the leading term, i.e. the term which has the largest exponential, so that

$$z_i(t) \approx \left(\prod_{m=1}^{i} \frac{b_m}{a_i - a_{m-1}}\right) e^{a_i t}.$$

Then type i is at its maximum near

$$t \approx t_i = \frac{1}{a_{i-1} - a_i} \left(\log \frac{b_i}{a_i - a_{i-1}} + \sum_{m=0}^{i-2} \log \frac{a_{i-1} - a_m}{a_i - a_m}\right). \tag{7.22}$$

In particular, after time t_n, the type x_n will dominate.

Multidimensional competition dynamics. Equations (7.5-7.11) represent two parallel mutation cascades, that is, two sets of quasispecies equations, coupled via the common fitness term, ϕ. In order to use the

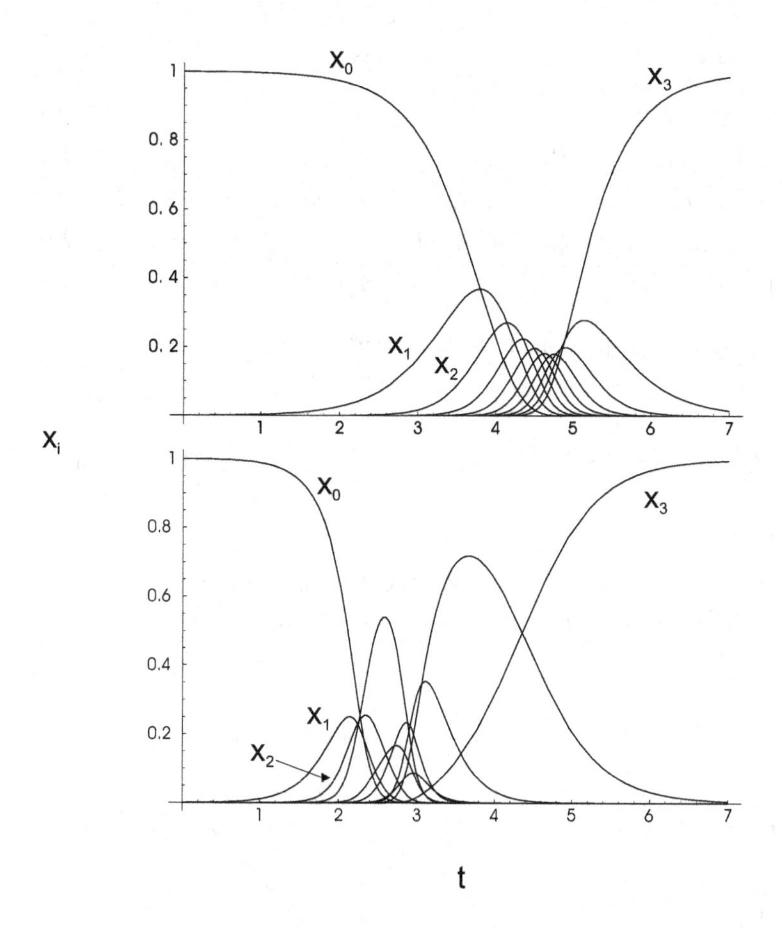

Fig. 7.3 Simulation of mutation cascades. In the picture above, we have $a_i = a_{i-1} + 2$, for $1 \leq i \leq n$ and $b_i = \epsilon$. In the picture below, we have $a_i = a_{i-1} + 2(1 + \xi_i)$ and $b_i = \epsilon(1 + \zeta_i)$, where ξ_i and ζ_i are some random numbers drawn from a uniform distribution between zero and one. For both pictures, $n = 9$, $c_i = 0$ for all i, $\epsilon = 0.001$ and $a_0 = 1$.

techniques developed above, let us write the equations for the mutational cascade in a simpler form,

$$\dot{z}_0 = a_0 z_0, \quad \dot{z}_i = b_i z_{i-1} + a_i z_i, \qquad 1 \leq i \leq n, \tag{7.23}$$

$$\dot{z}'_0 = a'_0 z'_0, \quad \dot{z}'_i = b'_i z'_{i-1} + a'_i z'_i, \qquad 1 \leq i \leq n, \tag{7.24}$$

$$\dot{z}_{n+1} = \sum_{k=0}^{n} (c_k z_k + c'_k z'_k), \tag{7.25}$$

by introducing the following obvious notations:

$$z_i \to S_i, \quad z_i' \to M_i, \quad 0 \le i \le n, \quad z_{n+1} \to w, \tag{7.26}$$

$$a_i = R_i(1-u_s), \quad a_i' = R_i(1-u_m), \quad 0 \le i \le n-1, \tag{7.27}$$

$$a_n = R_n[1-u_s+\alpha u(1-\epsilon_s)], \quad a_n' = R_n[1-u_m+\alpha u(1-\epsilon_m)], \tag{7.28}$$

$$b_i = \alpha u R_{i-1}(1-\epsilon_s), \quad b_i' = \alpha u R_{i-1}(1-\epsilon_m), \quad 1 \le i \le n, \tag{7.29}$$

$$c_i = (1-\alpha)R_i u(1-\epsilon_s), \quad c_i' = (1-\alpha)R_i u(1-\epsilon_m), \quad 0 \le i \le n. \tag{7.30}$$

In a matrix notation, equations (7.5-7.11) read:

$$\dot{\mathbf{z}} = \hat{f}_s \mathbf{z}, \quad \dot{\mathbf{z}}' = \hat{f}_m \mathbf{z}', \tag{7.31}$$

where the fitness matrices $\hat{f}_{s,m}$ are found from (7.23-7.25). The solution of the nonlinear system can be found by re-normalizing the solution of system (7.23-7.25), as before.

Fitness landscape. In order to analyze the dynamics of system (7.5-7.11), we have to make assumptions on the fitness landscape for the consecutive mutants (Figure 7.4).

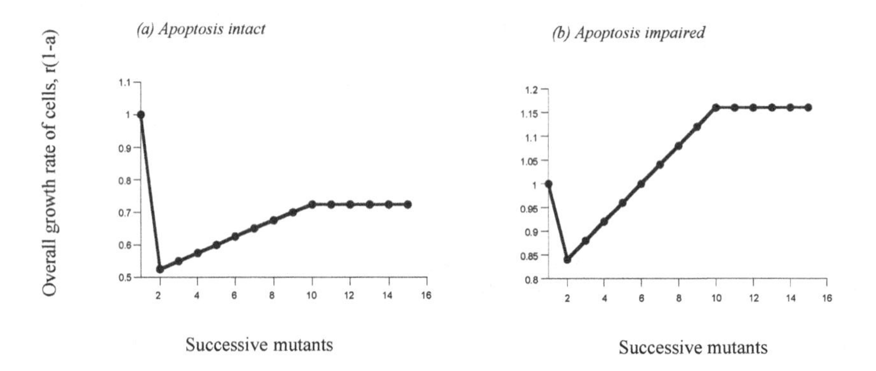

Fig. 7.4 Fitness landscape as a result of the successive accumulation of mutations by cells. We distinguish two scenarios. (a) If apoptosis is intact, accumulation of mutations results in a lower fitness compared to unaltered cells. Even if the mutations result in an increased rate of cell division, the induction of apoptosis in mutated cells prevents them from attaining a higher fitness than the unaltered cells. (b) If apoptosis is impaired, the accumulation of successive mutations will eventually result in a higher fitness compared to unaltered cells. The exact shapes of the curves are not essential. What is important is whether the mutants will eventually have a lower (a) or higher (b) intrinsic reproductive rate.

Since we are interested in cancer progression, we assume that the intrinsic rate of cell division of the consecutive mutants, r_i, increases ($r_{i+1} > r_i$). Such mutations could correspond to alterations in oncogenes or tumor suppressor genes. Because an accumulation of mutations cannot result in an infinite increase in the division rate of cells, we assume that the division rate plateaus. Once the cells have accumulated n mutations, we assume that further viable mutants are neutral because the division rate cannot be increased further. (This end stage of the mutational cascade is thus mathematically identical to the simple model discussed in the last section.) While we assume that the consecutive mutants can divide faster, they can also carry a disadvantage: the mutations can be recognized by the appropriate checkpoints which induce apoptosis. With this in mind, we will consider two basic types of fitness landscapes. If $r_0 > r_n(1-a)$, the intrinsic growth rate of the mutated cells, S_i and M_i, will be less than that of the unaltered cells, S_0 and M_0 (Figure 7.4). While the mutations allow the cells to escape growth control, the mutated cells are killed at a fast rate by apoptosis upon cell division. This scenario corresponds to the presence of efficient apoptotic mechanisms in cells. On the other hand, if $r_0 < r_n(1-a)$, the accumulation of mutations will eventually result in an intrinsic growth rate which is larger than that of unaltered cells (Figure 7.4). While mutated cells can still undergo apoptosis upon cell division, apoptosis is not strong enough to prevent an increase in the intrinsic growth rate. Hence, this scenario corresponds to impaired apoptosis in cells. In the following sections, we study the competition dynamics between stable and mutator cells in an evolutionary setting, assuming the presence of relatively strong and weak apoptotic responses.

Time scale separation. In what follows we will assume that the dynamics of the two cell populations happen on two different time scales. In other words, we require that the stable population is still in the state S_0 while the unstable population has already produced all mutants and reached a quasistationary state.

The typical time, t_1^s, of change for the type S_0 is found from equation (7.22). Similarly, we can find the time, t_n^m, it takes to reach the state M_n. It is given by the same equation (7.22) except the coefficients in the nth equation must be replaced by the corresponding coefficients with primes. The mapping to the biological parameters is found from (7.26-7.30). Note that using formula (7.22) has its restrictions, and in the case where it is not applicable, one can directly calculate t_n^m by estimating the time it takes for

z_n to reach its maximum (see formula (7.21)). The inequality

$$t_1^s \ll t_n^m \tag{7.32}$$

guarantees that by the time the unstable population has traveled down the mutation cascade to approach its quasistationary distribution, the stable population of cells is still dominated by S_0.

The conditions for the reversal of competition. In the multidimensional competition problem, equations (7.31), the outcome is determined by the largest eigenvalue of the fitness matrices, $\hat{f}_s$ and $\hat{f}_m$. As time goes by, the unstable cell population will approach its stationary distribution (defined by the eigenvector corresponding to the principal eigenvalue), and its fitness is given by the eigenvalue. Because of the time-scale separation, we will assume that during this time, the stable population remains largely at the state S_0. Thus the "winner" of the competition is defined by comparing the two eigenvalues, a_0 and a'_n, see equations (7.27-7.28).

Let us define the value of u, u_c, so that for $u = u_c$, we have $a_0 = a'_n$. As the hit rate passes through u_c, the result of the competition reverses. We have

$$u_c = \frac{R_0 - R_n}{R_0 - R_n - \beta(R_0\epsilon_s - R_n\epsilon_m) + R_n\alpha(1 - \epsilon_m)}.$$

In order to determine whether competition reversal takes place for each scenario (see below), we need to make sure that the following condition is satisfied:

$$0 \le u_c \le 1.$$

In the next sections we will examine different parameter regimes and conclude that competition reversal may or may not take place; we derive the exact conditions for this. In what follows, we will use several definitions. Let us set

$$\Delta\epsilon = \epsilon_s - \epsilon_m,$$

and denote by $\bar{\epsilon}_s$ the following threshold value of ϵ_s,

$$\bar{\epsilon}_s = \frac{\alpha R_n}{\beta R_0 + R_n(\alpha - \beta)}. \tag{7.33}$$

This quantity is defined from setting $u_c = 1$ and $\Delta\epsilon = 0$. Finally, we define the critical *gap*, $\Delta^*\epsilon$, between the two values of ϵ, by setting $u_c = 1$:

$$\Delta^*\epsilon = \epsilon_s \left[1 + \frac{\beta R_0}{(\alpha - \beta)R_n}\right] - \frac{\alpha}{\alpha - \beta}.$$

Now, let us go back to the two types of fitness landscape, Figure 7.4, and examine the scenarios of strong and weak apoptosis separately.

7.2.2 *Strong apoptosis*

Here we assume that the apoptotic mechanisms in cells are strong. That is, $r_0 > r_n(1 - a)$ (Figure 7.4a). This means that although the successive mutations will allow the cell to divide faster, the induction of apoptosis ensures that the intrinsic growth rate of the mutants is lower compared to unaltered cells. Note that it is not necessary to assume that oncogenic mutations allow cells to divide faster. Indeed, some cancer cells may progress more slowly through the cell cycle than healthy cells. The important assumption is that accumulation of mutations lowers the intrinsic growth rate of the cells.

In this scenario, the intrinsic growth rate of the stable cells, S, is higher than that of the unstable cells, M. The reason is as follows. The population of stable cells, S, has efficient repair mechanisms. Thus, most cells will remain at the unaltered stage, S_0. Because population M is unstable, a higher fraction of this cell population will contain mutations. Since these mutations impair reproduction (e.g. because of induction of apoptosis), the intrinsic growth rate of the unstable cells, M, is lower than that of the stable population, S.

At low DNA hit rates, the cells with the faster intrinsic growth rate win the competition. Thus, at low DNA hit rates (low value of u), the stable phenotype, S, wins (Figure 7.5a). On the other hand, at higher DNA hit rates (high value of u), the outcome of competition can be reversed because frequent cell cycle arrest significantly reduces growth. That is, the genetically unstable cells, M, may win and take over the population. As in the simple model discussed above, it requires that the cost of cell cycle arrest is higher than the cost associated with the generation of deleterious mutants (i.e. $C_{arr} > C_{del}$). Furthermore, reversal of competition may require that the repair rate of stable cells (ϵ_s) lies below a threshold, and that there is a sufficient difference in the repair rate between stable and unstable cells.

As the population of unstable cells wins, they accumulate mutations. Even if the sequential mutants are disadvantageous because of the induction of apoptosis, the high mutation rate pushes the population down the mutational cascade. While all variants, M_i, persist, the distribution of the variants becomes skewed toward M_n as the DNA hit rate is increased.

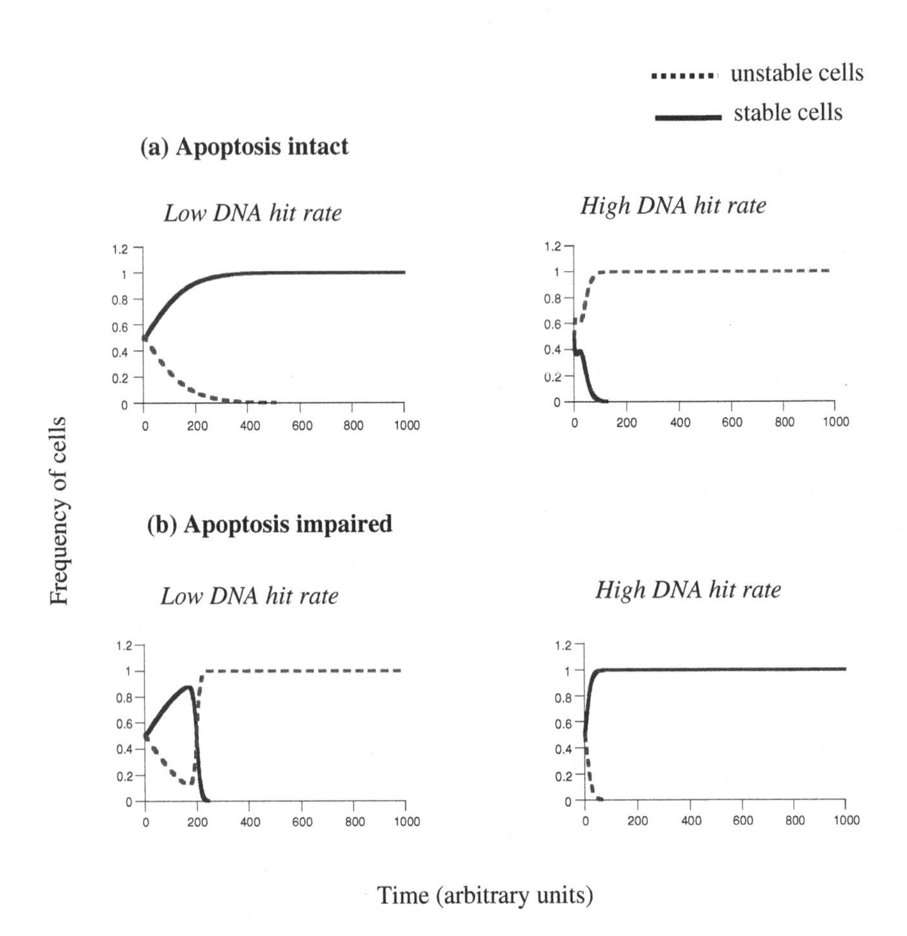

Fig. 7.5 DNA damage and the selection of genetic instability. (a) Cells have intact apoptotic responses. At low DNA hit rates stability wins. At high DNA hit rates instability wins. (b) Cells have impaired apoptotic responses. At low DNA hit rates, instability wins. At high DNA hit rates stability wins. Parameter values were chosen as follows: $\epsilon_s = 0.99$; $\epsilon_m = 0.1$; $\beta = 0.2$. For (a) $\alpha = 0.6l$ $a = 0.5$. For (b) $\alpha = 0.1$; $a = 0.2$. Low DNA hit rate corresponds to $u = 0.07$, and high DNA hit rate corresponds to $u = 0.7$. Fitness landscapes for successive mutants are given in Figure 7.4.

These results can be obtained by a very simple analysis of the relative values of the relevant eigenvectors, see (7.27-7.28), and by finding conditions under which reversal can occur. In mathematical terms, strong apoptosis corresponds to the situation where

$$R_0 > R_n.$$

From definition (7.33), $0 \leq \bar{\epsilon}_s \leq 1$. Also, we will use the fact that $\Delta^*\epsilon$ grows with ϵ_s, so that $\Delta^*\epsilon > 0$ for $\epsilon_s > \bar{\epsilon}_s$ and $\Delta^*\epsilon < 0$ for $\epsilon_s < \bar{\epsilon}_s$. We can distinguish the following two cases:

- If $\beta > \alpha$ (which is the same as $C_{arr} < C_{del}$), we have $a_n > a'_n$, which means that M_n *never* corresponds to the largest eigenvalue. This means that the stable cells always win and the competition reversal does not happen. (Technically speaking, the reversal happens between a_0 and a_n rather than a_0 and a'_n.)
- If $\beta < \alpha$ (which is the same as $C_{arr} > C_{del}$), then competition reversal will happen if the following condition is satisfied: $\Delta\epsilon > \Delta^*\epsilon$ (this is because the function u_c decays with $\Delta\epsilon$). We also observe that in this case, $\Delta^*\epsilon$ is a growing function of ϵ_s, which reaches zero at $\epsilon_s = \bar{\epsilon}$, with $0 \leq \bar{\epsilon} \leq 1$. We have two subcases:

  (a) For $\epsilon_s < \bar{\epsilon}_s$, we have $\Delta^*\epsilon < 0$, and the reversal happens for *any* difference between ϵ_s and ϵ_m.

  (b) For $\epsilon_s > \bar{\epsilon}_s$, $\Delta^*\epsilon > 0$, and we need a finite gap between ϵ_s and ϵ_m, $\Delta\epsilon > \Delta^*\epsilon$. We also have to make sure that $\Delta^*\epsilon < \epsilon_s$, which gives the condition

$$\epsilon_s < \frac{\alpha R_n}{\beta R_0}.$$

The biological interpretation of these conditions was given in the beginning of this section.

These results have important practical implications. The model tells us that in the presence of intact apoptotic mechanisms, a high DNA hit rate selects in favor of genetic instability, while the tissue remains stable and unaltered if the DNA hit rate is low. A high DNA hit rate can be brought about both by the presence of carcinogens, or by chemotherapy. Therefore, if healthy tissue is exposed to carcinogens, we expect genetic instability to rapidly emerge and this can give rise to cancer progression. In the same way, chemotherapy can select for genetic instability in otherwise healthy tissue and thus induce new tumors as a side effect.

7.2.3 Weak apoptosis

Now we assume that the apoptotic mechanisms in cells are impaired. That is, $r_0 < r_N(1-a)$, Figure 7.4b. This means that accumulation of mutations will eventually result in the generation of variants which have a faster intrinsic growth rate compared to unaltered cells. Thus, in principle, both populations are expected to eventually evolve toward the accumulation of mutations and progression to cancer. Hence, both stable and unstable cancers can be observed. However, as we noted before, we assume that these processes occur over different time scales for the two populations of cells, condition (7.32).

If the stable and unstable populations compete, the unstable population will have a higher intrinsic growth rate than the stable population (because the induction of apoptosis in response to mutation is inefficient). Therefore, at low DNA hit rates, u, the mutator phenotype, M, wins the competition (Figure 7.5). If the DNA hit rate is increased, the competition can be reversed in favor of the stable cell population, S. This requires that the cost of generating deleterious mutants be greater than the cost of cell cycle arrest (i.e. $C_{del} > C_{arr}$). Furthermore, a sufficient difference in the repair rate of stable and unstable cells is required to reverse the outcome of competition.

Here is the reasoning behind these conclusions. For weak apoptosis, we have

$$R_0 < R_n.$$

In this case, if $(R_n - R_0)/R_n < \alpha/\beta$, then $\bar{\epsilon}_s > 1$. If on the other hand $(R_n - R_0)/R_n > \alpha/\beta$, then $\bar{\epsilon}_s < 0$. We have the following two cases:

- If $\beta > \alpha$, then the function u_c decays with $\Delta\epsilon$, so for reversal to occur we need to have $\Delta\epsilon > \Delta^*\epsilon$.
  (a) For $(R_n - R_0)/R_n < \alpha/\beta$, the function $\Delta^*\epsilon$ decays with ϵ_s and crosses zero at $\epsilon_s = \bar{\epsilon} > 1$. This means that for all ϵ_s, $\Delta^*\epsilon > 0$. We need to require that $\Delta^*\epsilon < \epsilon_s$, which gives the condition

$$\epsilon_s > \frac{\alpha R_n}{\beta R_0}.$$

  If this condition holds then the reversal occurs, as long as $\Delta\epsilon > \Delta^*\epsilon$; that is, the difference in repair rates must be larger than the critical value.
  (b) For $(R_n - R_0)/R_n > \alpha/\beta$, the function $\Delta^*\epsilon$ grows with ϵ_s and crosses zero at $\epsilon_s = \bar{\epsilon} < 0$. This means that for all ϵ_s, $\Delta^*\epsilon > 0$.

We need to require that $\Delta^*\epsilon < \epsilon_s$, which gives again the condition

$$\epsilon_s > \frac{\alpha R_n}{\beta R_0}.$$

If this condition holds then the reversal occurs as long as $\Delta\epsilon > \Delta^*\epsilon$.

- If $\beta < \alpha$, then the function u_c grows with $\Delta\epsilon$, so we need to have $\Delta\epsilon < \Delta^*\epsilon$.

  (a) Condition $(R_n - R_0)/R_n > \alpha/\beta$ is impossible to satisfy, so reversal does not happen in this case.
  (b) If $(R_n - R_0)/R_n < \alpha/\beta$ then $\Delta^*\epsilon$ is a growing function of ϵ_s which crosses zero at $\epsilon_s = \bar{\epsilon} > 1$. This means that for all ϵ_s, $\Delta^*\epsilon < 0$, and reversal is again impossible.

Our results have practical implications. If cells develop a mutation resulting in impaired apoptotic responses, then genetic instability has a selective advantage if the DNA hit rate is low. Therefore, even if there is no exposure to carcinogens, a chance loss of apoptosis can result in the outgrowth of genetic instability and thus progression of cancer. On the other hand, if there is a growing cancer with impaired apoptotic responses, our results suggest that an elevation of the DNA hit rate by chemotherapeutic agents can reverse the relative fitness in favor of stable cells, and this can result in cancer reduction or slower progression.

A note of clarification: in the above arguments we assumed for simplicity that apoptosis is inefficient in both the unstable and the stable cells. The arguments about chemotherapy, however, remain robust even if we assume that only the mutator phenotypes have impaired apoptosis, while the stable and healthy population of cells has intact apoptotic responses. The reason is that over the time frame considered, the population of stable cells remains genetically unaltered (i.e. at stage S_0). Since the cells are unaltered, the presence or absence of apoptosis does not change the dynamics.

7.3 Summary of mathematical results

The equations have examined the competition dynamics between genetically stable and unstable populations of cells. They identified under which circumstances genetic instability is selected for or against in the context of cancer progression. In particular, they examined the role of the rate at which DNA is damaged.

Table 7.2 Summary of the results gained from the model which takes into account evolution and mutation cascades. If apoptosis is intact, mutators (M) have a lower intrinsic growth rate than stable cells (S). Hence, a high DNA hit rate can select for M. If apoptosis is impaired, M have a higher overall intrinsic growth rate than S. Thus, a high DNA hit rate can select in favor of S.

	Apoptosis intact	**Apoptosis impaired**
Low DNA hit rate	S win	M win
High DNA hit rate	M win if $C_{arr} > C_{del}$	S win if $C_{arr} < C_{del}$

A change in the DNA hit rate can reverse the outcome of competition. In the simplest setting, an increase in the DNA hit rate can switch the outcome of competition in favor of cells characterized by a slower intrinsic growth rate. This requires a sufficient difference in the repair rate between the stable and mutator cells, and a condition on the relative values of costs associated with cell cycle arrest and creation of deleterious mutants. The conditions under which genetic instability is selected for depends on the efficacy of apoptosis. In terms of cancer evolution and progression, this gave rise to the following insights (Table 7.2).

- If apoptosis is strong, accumulation of mutations by unstable cells slows down the intrinsic growth rate because of the frequent induction of cell death. Thus, stable cells have a higher intrinsic growth rate than mutators. Consequently, at low DNA hit rates, the stable cells win. The presence of high DNA hit rates can, however, result in the selection and emergence of the genetically unstable cells. This occurs if the cost of cell cycle arrest upon repair is higher than the cost of creating deleterious mutations.
- On the other hand, if apoptotic responses are impaired, accumulation of mutations by unstable cells will not result in frequent cell death upon division. Therefore, the intrinsic growth rate of unstable cells can be higher than that of stable cells if adaptive mutations are acquired. In this case, genetic instability is expected to emerge at low DNA hit rates. At high DNA hit rates, however, genetic instability can be selected against and mutators can go extinct. This occurs if the cost of creating deleterious mutations is higher than the cost of cell cycle arrest.

7.4 Selection for genetic instability

A fascinating question is how much genetic instability can contribute to faster adaptation and evolution of cancer cells [Jackson and Loeb (1998); Jackson and Loeb (2001); Loeb and Loeb (2000); Loeb (1991); Loeb (2001); Tomlinson (2000); Tomlinson and Bodmer (1999); Tomlinson *et al.* (1996)]. It can be argued that genetic instability can be selected for due to the following two factors:

(i) Genetic instability can be advantageous if it results in a faster accumulation of adaptive mutations compared to stable cells [Loeb (1991)]. This could allow the cancer to evolve faster and to overcome selective barriers and host defenses. An example are tumor suppressor genes where both copies have to be mutated. Instead of the occurrence of two independent point mutations, loss of heterozygocity in genetically unstable cells can result in the loss of suppressor function if one copy has been mutated.

(ii) Genetic instability can be advantageous simply because cells avoid delay in reproduction associated with repairing and maintaining the genome [Breivik and Gaudernack (1999a); Breivik and Gaudernack (1999b)]. When genetic damage is detected, the relevant checkpoints induce cell cycle arrest during which the damage is repaired. If genetic damage occurs often, frequent arrest significantly slows down the replication rate of the cells, and loss of repair can be advantageous. Experimental evidence supports this notion. Bardelli et al. [Bardelli *et al.* (2001)] have shown that exposure to specific carcinogens can result in the loss of the checkpoint that was induced by the carcinogenic agent used.

At this stage, it is unclear what selective mechanism is responsible for the emergence of genetic instability (or in fact whether genetic instability appears simply as a side effect of other genetic alterations on the way to cancer). It is possible that different types of genetic instability can have different effects on the evolution of the cell populations. The increased rate at which the quasispecies travels up the fitness landscape may or may not be out-weighed by the costs associated with creating deleterious mutations. This in turn may depend on the nature of the instability. In particular, it may be determined by whether the genetic changes are relatively small (such as in MIN) or larger (such as in CIN, see Chapter 6).

If the main driving force for the emergence of genetic instability is avoidance of cell cycle arrest (rather than faster adaptation), this could contribute to explaining why certain instabilities are specific to certain types of cancers or tissues. Different environments can cause different types of genetic alterations which induce separate checkpoints [Bardelli *et al.* (2001)]. The checkpoints which are lost in the cancer would be the ones which are most often induced in the appropriate environment and tissue surroundings. On the other hand, if genetic instability mainly emerges because it allows the cells to adapt faster, we expect that instability is lost at later stages of cancer progression. This is because the cancer has evolved to an optimal phenotype, and now stability avoids deleterious mutations and thus increases fitness [Cahill *et al.* (1999)].

7.5 Genetic instability and apoptosis

If genetic instability can result in a faster accumulation of adaptive mutations (case (i) above), it could in principle be the driving force of cancer progression. As pointed out in the previous section, it is unclear whether this is the case, or whether alternative selection pressures are responsible for the emergence of genetic instability. Here, we assume that instability can result in the accumulation of adaptive mutations and explore possible pathways to the emergence of genetic instability and cancer progression. Assume we start from a wild-type cell which is stable and has intact apoptotic mechanisms. The mathematical model suggests that genetic instability can only drive progression toward fitter phenotypes if apoptosis is impaired. This is because in the presence of intact apoptosis, accumulation of mutations results in elevated levels of cell death which slows down the intrinsic growth rates. Thus, to gain a fitness advantage, both apoptosis, and stability genes have to be mutated. This can occur via two pathways (Figure 7.6).

(1) In the first pathway, the initial mutation impairs the apoptotic response in the cell. This variant is selectively neutral compared to the wild-type. The reason is that the cell still has intact repair mechanisms. Therefore, mutations are unlikely to be generated in the time frame considered. As long as mutations do not accumulate, the presence or absence of apoptotic mechanisms does not change the dynamics of cell growth. Following this mutation, a second mutation is generated which

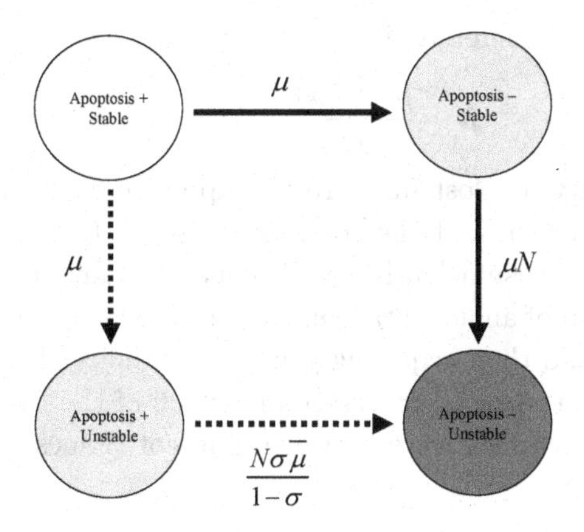

Fig. 7.6 Pathways toward the selection of genetic instability and cancer progression.

confers genetic instability. This mutant will be selected for instantly.

(2) In the second pathway, the initial mutation confers genetic instability to the cell. Since apoptotic responses are still intact, the model analysis tells us that this variant will have a lower fitness compared to the wild type and will be on its way to extinction. However, because the cell is unstable, it can generate mutations at a faster rate. Thus, there is a chance that the mutation impairing apoptosis is generated before this cell variant has gone extinct. As soon as the apoptotic mechanism has been impaired, the unstable cell gains a selective advantage.

We can calculate which of these two pathways occurs faster, and this is the pathway that is more likely to lead to selection of instability (for mathematical details of this approach, see Chapters 2 and 4). We introduce the following notation (Figure 7.6). The number of cells in a tissue is denoted by N. The rate at which a genetically stable cell mutates (to be either unstable or apoptosis impaired) is given by μ. The rate at which an unstable cell mutates toward a loss of apoptotic function is denoted by $\bar{\mu}$. Thus, $\bar{\mu} > \mu$. The relative reproductive rate of an unstable cell which has intact apoptotic responses is given by $\sigma < 1$ (we assume that the wild-type reproductive rate is 1), which reflects the fact that unstable cells with intact apoptosis have a selective disadvantage. The rate at which an advantageous mutator is generated via the first pathway (first *apoptosis*$^-$, then *repair*$^-$) is given by $\mu^2 N$. The rate at which an advantageous mutator is generated via the second pathway (first *repair*$^-$, then *apoptosis*$^-$) is given by $N\mu\sigma\bar{\mu}/(1-\sigma)$. Therefore, if

$$\bar{\mu} > \frac{\mu(1-\sigma)}{\sigma},$$

then repair and stability are lost first. In the opposite case, apoptosis is lost first. In biological terms, if the relative fitness of the unstable and apoptosis competent cell is sufficiently low (because mutants are killed efficiently), then generation of an advantageous mutator is likely to proceed by first losing apoptosis, and then acquiring genetic instability. Knowledge of parameter values will be required to determine which of the two pathways is more likely. The result might vary between different tissues.

7.6 Can competition be reversed by chemotherapy?

The results derived in this chapter have implications for the use of chemotherapy (Figure 7.7). Chemotherapy essentially increases the degree of DNA damage. Therefore, it can be used to reverse the relative fitness of stable and unstable cells such that unstable cells are excluded (Figure 7.7). This can drive progressing cancer cells extinct and result in the persistence of stable cells. These may either be healthy cells or less aggressive and slowly progressing tumor cells. In order to achieve this outcome, there needs to be a sufficient difference in the repair rate between stable and unstable cells. If this is not the case, therapy can merely slow down cancer progression.

Since in this scenario, chemotherapy acts by modulating the competition between stable and unstable cells, it is not a requirement that every last cell is killed by the drugs. Selection and competition will make sure that the unstable cancer cells are driven extinct. This argument, however, requires that there is an element of competition between unstable and stable cells. Whether and under which circumstances this is the case remains to be determined.

This is a different mechanism of action compared to the traditional view which assumes that chemotherapy only acts by killing dividing cells. For chemotherapy to reverse the fitness of stable and unstable cells, two conditions are required. (i) There needs to be a sufficient difference in the repair rate between stable and unstable cells. The higher the replication rate of unstable cells relative to stable cells, the higher this difference in the repair rate required to achieve success. Therefore, contrary to traditional views, a faster rate of cell division of cancerous unstable cells renders successful

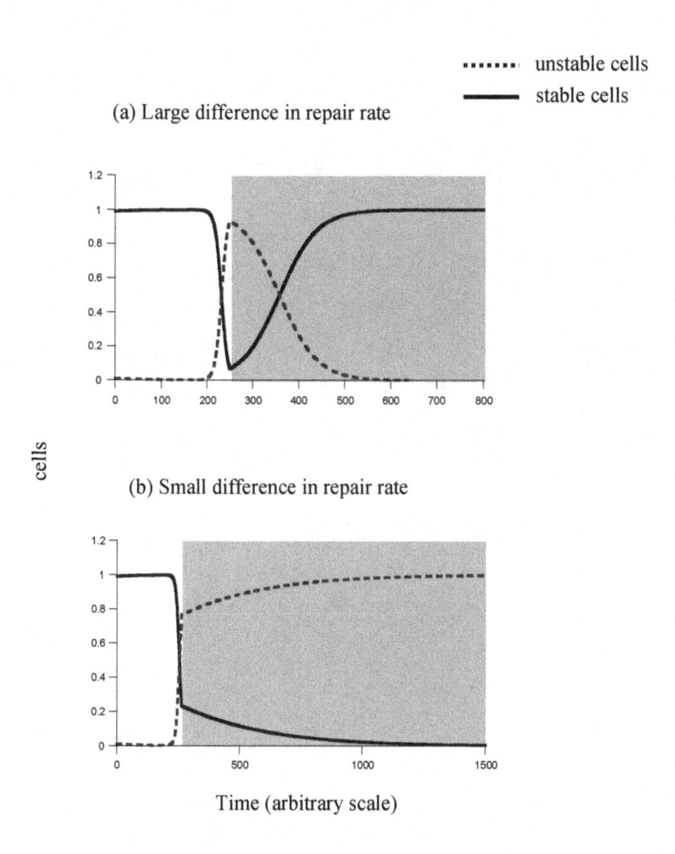

Fig. 7.7 Simulation of chemotherapy, modeled by an increase in the DNA hit rate, u. We start with a situation where cells which are unstable and have impaired apoptosis spontaneously give rise to cancer growth and progression. (a) If there is a large difference in repair rate between stable and unstable cells, therapy can exclude the unstable cells. (b) If there is a smaller difference between stable and unstable cells, then therapy fails to exclude instability. Parameters were chosen as follows: $\epsilon_s = 0.99$; $\beta = 0.2$; $\alpha = 0.1$. We assume that the degree of apoptosis differs between stable and unstable cells. Stable cells have intact apoptosis ($a = 0.5$), while unstable cells have impaired apoptosis ($a = 0.2$). For (a) $\epsilon_m = 0.1$. For (b) $\epsilon_m = 0.4$. Low DNA hit rate corresponds to $u = 0.07$, and high DNA hit rate corresponds to $u = 0.8$. Fitness landscapes for successive mutants are given in Figure 7.4.

treatment more difficult in this scenario. (ii) The cost of generating lethal mutants in unstable cells must be higher than the cost of cell cycle arrest in stable cells. If this is not the case, it does not pay to retain repair mechanisms, and the fitness of unstable cells can never be reversed. In this case, treatment has a higher negative impact on stable than on unstable cells, and the mutators are resistant.

Chapter 8

Tissue aging and the development of cancer

Chapter 1 discussed cellular mechanisms which prevent cells from acquiring genomic alterations and from becoming cancerous. When cells are damaged, so called *checkpoint genes* induce repair, senescence, or apoptosis [Chavez-Reyes *et al.* (2003); Itahana *et al.* (2001); Offer *et al.* (2002)]. The most prominent checkpoint gene is p53. It is thought to play a role in all three responses [Vogelstein *et al.* (2000a)] and is mutated in more than half of all human cancers. Loss of p53 is a defining event in the progression to malignancy in many cases. There are many other cellular checkpoints which have a similar role.

The activity of checkpoints such as p53 can, however, also have a different effect. If tissue cells become damaged very often, cells might frequently undergo apoptosis or enter permanent senescence. This could reduce the size of the tissue and the ability of the tissue to function in a healthy way. The amount of DNA damage increases with age [Finkel and Holbrook (2000)]. The reason is that metabolism generates reactive oxygen species as a byproduct, and this has a genotoxic effect. Since damage can induce p53 activity, aging can manifest itself in increased amounts of cell death or senescence. Therefore, high levels of checkpoint activity, as a consequence of aging-induced genotoxic events, could contribute to the symptoms of aging in mammals or humans [Campisi (2003a); Hasty *et al.* (2003); Oren (2003); Schmitt (2003)].

Therefore, there seems to be a tradeoff in the presence of high DNA damage. On the one hand, checkpoint activity prevents the accumulation of mutations and the development of cancer. On the other hand, it promotes aging which in itself can lead to morbidity and mortality. However, the relationship between checkpoint competence, the development of cancer, and the development of age related symptoms can be rather intricate.

In addition to the simple tradeoff mentioned above, there are additional complicating factors. For example, healthy tissue cells might play a crucial role in exerting inhibitory effects on cancer cells and in preventing the development of cancer [Hsu *et al.* (2002); Mueller and Fusenig (2002)]. Such interactions between cancer and the microenvironment were described in Chapter 1. In this case, increased checkpoint activity could correlate with a higher chance to develop cancer. The reason is that increased checkpoint activity results in tissue aging, and this in turn results in less tumor inhibition.

This chapter explores possible connections between checkpoint competence, aging, and the development of cancer. How does checkpoint activity correlate with the life span of an organism? We will consider organisms which vary in their checkpoint competence. That is, we will compare genotypes which are more or less efficient at detecting and reacting to DNA damage. We will relate this to mice which vary in their p53 competence. At one end of the spectrum, there are p53 deficient mice (p53-/-). All cells of these mice lack functional p53. Then, there are p53+/- and p53+/+ mice. The p53 +/- mice have one functional, and one inactivated copy of p53. The p53+/+ mice have two functional copies of p53. Both p53 +/- and p53+/+ mice are thought to be wild type, because one copy of the gene is sufficient to display normal function, although this notion is still under debate. At the other end of the spectrum, there are so–called super p53 mice. These have additional copies of p53 that are active in all cells of the animal. These organisms show an increased ability to detect and react to DNA damage.

8.1 What is aging?

It is important to define the concept of aging in the current context. Aging is a complex phenomenon and involves many different aspects. As humans approach middle age, the functions of the body start to deteriorate. There is a general decline in physical function, and possibly in mental function as well. As humans age, they become more prone to a variety of diseases. These involve, for example, the cardiovascular, immune, nervous, digestive, and urinary systems. Prevalent diseases associated with aging include heart attacks, Alzheimer's disease, arthritis, diabetes, and cancer. Most diseases which are associated with aging might have a similar underlying genetic and cellular cause. The diseases are generally thought to be linked to cellular senescence [Campisi (2001); Campisi (2003b)]. Cellular senescence is defined as the occurrence of infinite cell cycle arrest. That is, although the cells do not die, they stop dividing and remain in a dormant state.

As mentioned briefly in the introduction, cellular senescence can be caused by the very checkpoint genes which are responsible for preventing cancer. In particular, senescence is induced if a cell experiences a high amount of DNA damage which cannot be repaired anymore. As explained above, the amount of DNA damage which cells experience increases with age as a result of the production of reactive oxygen species in metabolism [Finkel and Holbrook (2000)]. This can lead to increased amounts of cellular senescence in tissues and can lead to a decline in tissue function. In the context of this chapter, we will therefore give the following definition of aging: *Aging is the decline in tissue function as a result of high levels of cellular senescence which occurs in the face of elevated DNA damage.* On a more physical level this can translate into the weight of an organism or of specific organs. Such criteria of aging have been used in experiments with mice which examined the relationship between p53 and aging [Tyner *et al.* (2002)]. We will discuss this extensively in this chapter.

8.2 Basic modeling assumptions

We consider a mathematical model which takes into account both the population of healthy tissue cells, and the population of first stage tumor cells [Wodarz (2004)]. The basic difference between these two is that the population size of healthy cells is constant, while the population of tumor cells can clonally expand. The central concept of the model is as follows. It starts with healthy tissue which experiences a certain amount of genome damage. This is either repaired or not depending on checkpoint competence. In the absence of repair, and if the mutation does not result in apoptosis, a transformed cell can be generated with a certain probability (first stage tumor cell). This tumor cell can expand and grow to a certain limited size. The model examines the characteristic time it takes for the number of tumor cells to increase to a defined value. The model does not consider growth of the tumor beyond this first stage. It does, however, examine the rate with which this first stage tumor can acquire additional mutations; this is a measure for the ability of the tumor to progress towards more advanced stages. The model is explained schematically in Figure 8.1. The general question which we ask is as follows. How does a reduction in checkpoint competence influence the initiation and progression of tumors?

We will study this question by numerical simulations. We determine how long it takes the cancer to grow to a defined (arbitrary) size threshold. This can be done with any standard ODE solver, or by writing a computer program which solves ordinary differential equations (using for example the Runge-Kutta method).

8.3 Modeling healthy tissue

We start with the population of healthy cells. We distinguish between two sub-populations: functioning cells, x, and arrested cells, y. When a cell divides, the DNA of the daughter cells is damaged with a probability u which correlates with the application of carcinogenic agents. With a probability η this damage results in cell cycle arrest (Figure 8.1). The arrested cell repairs the DNA damage and returns to a functioning state at a rate γ. The lower the value of γ, the longer the duration of cell cycle arrest. If $\gamma=0$, there is indefinite senescence. With a probability $1-\eta$, the DNA damage does not result in cell cycle arrest and repair. Now there are two possibilities. With a probability α, the damage leads to cell death, e.g. as a result of apoptosis. With a probability $1-\alpha$, the damaged or mutated cell survives. With a probability β the mutation is carcinogenic; that is, a transformed cell is generated. With a probability $1-\beta$ the mutation does not contribute to the process of carcinogenesis. For simplicity it is assumed that these cells are selectively neutral.

Therefore, checkpoint competence is captured in two parameters. The parameter η describes the probability that damage will result in cell cycle arrest and repair. The parameter α describes the probability that damage results in apoptosis. The value of these parameters can range between zero and one. A lower value of η corresponds to a lower probability of arrest and repair in all cells and could in practical terms correspond to e.g. a p53 knockout organism . If $\eta=0$, damage is never repaired. If $\eta=1$, repair always happens upon damage and mutations can never occur. The same applies to the probability to undergo apoptosis upon DNA damage.

These dynamics are given by the following set of differential equations which describe the development of the cell populations over time.

(a)

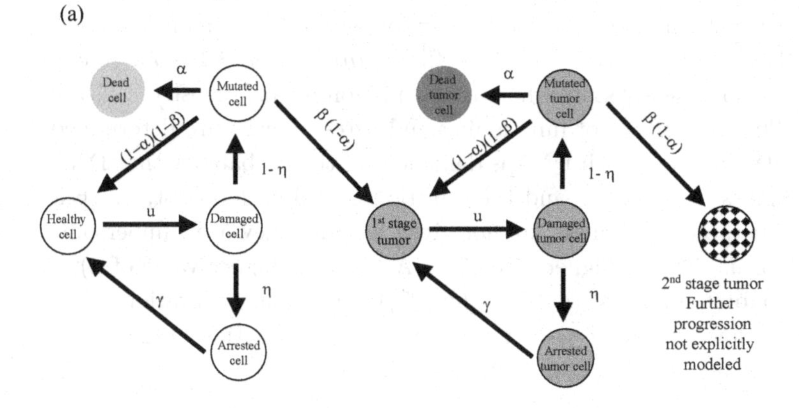

(b)

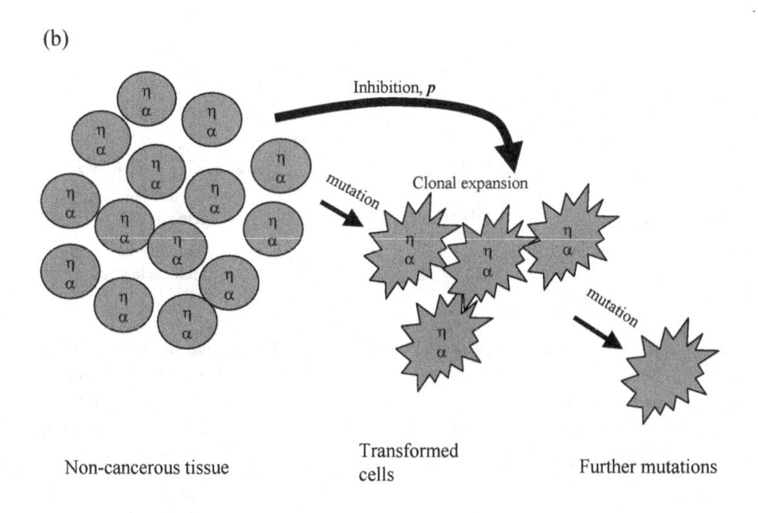

Fig. 8.1 Schematic representation of the principles which underlie the mathematical model. (a) Basic cellular processes in healthy and tumor cells. (b) Dynamics which underlie the equations.

$$\dot{x} = \rho x \left[1 - u + u\left(1 - \eta\right)\left(1 - \alpha\right)\left(1 - \beta\right)\right] + \gamma y - \phi x,$$
$$\dot{y} = \eta u \rho x - \gamma y - \phi y.$$

The parameter ρ denotes the turnover rate of the cells, and ϕ is an expression which ensures that the population size $x + y$ remains constant ($\phi = \rho x\left[1 - u + u\left(1 - \eta\right)\left(1 - \alpha\right)\left(1 - \beta\right)\right] + \eta u \rho x$; i.e. $y = 1 - x$). These dynamics go to a steady state where both functioning and arrested cells are present. The frequencies of functioning and arrested cells are determined mainly by the rate at which DNA is damaged, u, the probability that DNA damage results in arrest, η, and the duration of cell cycle arrest, γ. (Expressions for these frequencies are not given because they are cumbersome and un-informative.) A higher rate of DNA damage (higher value of u), a higher probability of arrest (higher value of η), and a longer duration of cell cycle arrest (lower value of γ) corresponds to a lower fraction of functional cells in the steady state (Figure 8.2).

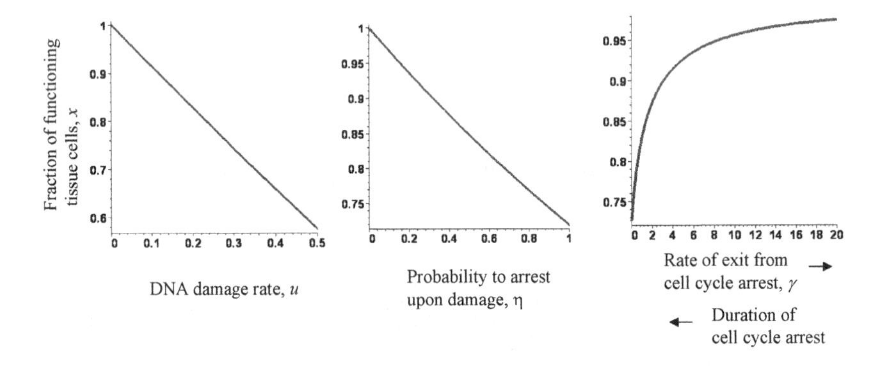

Fig. 8.2 The fraction of functional tissue cells is a function of the rate at which DNA becomes damaged (u), checkpoint competence (the probability to arrest upon damage, η) and the duration of cell cycle arrest (expressed as the rate at which cells exit from the arrested state, γ). Parameter values were chosen as follows: ρ=2; γ=0.1; u=0.3; η=0.9; α=0.5; $\beta=10^{-7}$; $d = 0.1$.

The effect of apoptosis on the number of tissue cells is difficult to determine. In the basic model as it is written here, tissue size is assumed to remain constant through homeostatic regulation. If cells die, the tissue compensates by producing more cells. Therefore, apoptosis does not have an effect on tissue size in this setting. During aging, however, it is likely that the ability of the tissue to renew itself becomes compromised. In this context, increased occurrence of apoptosis upon DNA damage can lead to a reduction in the number of functioning tissue cells, in the same manner as shown for the induction of cell cycle arrest. Because the relationship between aging and apoptosis is currently unclear [Campisi (2003a);

Camplejohn *et al.* (2003); Gilhar *et al.* (2004); Jansen-Durr (2002)], and because the processes of cell renewal and proliferation upon death may also vary among different tissues, the effect of apoptosis will not be analyzed explicitly. Instead, the analysis concentrates on how the induction of cellular senescence influences the relationship between a decline in tissue function and the emergence of cancer [Campisi (2001); Parrinello *et al.* (2003)].

8.4 Modeling tumor cell growth

Next, consider the population of tumor cells. In addition to the basic cellular processes which have been described in the context of healthy cells, the tumor cell dynamics are characterized by some important additional features. The tumor cells escape homeostatic regulation and undergo clonal expansion rather than staying at a constant population size. In addition, the proliferation of tumor cells can be inhibited by the tissue microenvironment. This is a well documented process [Hsu *et al.* (2002); Mueller and Fusenig (2002)] and is explained in more detail before the model is constructed and analyzed.

Experiments strongly suggest that the genetic and molecular events which occur in cancerous cells are not sufficient to account for the process of carcinogenesis. The microenvironment in which the cancer develops may be equally important for the tumor cells to escape homeostatic control and to give rise to disease [Cunha and Matrisian (2002); Hsu *et al.* (2002); Tlsty (2001); Tlsty and Hein (2001)]. The stroma surrounding the tumors shows in many cases changes in the patterns of gene expression, cellular composition and the extracellular matrix. This allows cancers to grow and progress. The development of cancer can thus be seen as a conspiracy between tumor cells and their altered environment which allows uncontrolled growth. Under non-pathogenic conditions, the tissue environment can prevent tumor cells from growing to significant levels. Autopsies have revealed that multiple small and non-pathogenic tumors exist which have failed to progress [Folkman and Kalluri (2004)]. A major regulatory force in this context is the inhibition of angiogenesis, i.e. the formation of new blood supply which tumor cells need to proliferate [Bayko *et al.* (1998); Folkman (2002); Hahnfeldt *et al.* (1999a)]. The concept of angiogenesis was briefly reviewed in Chapter 1 and will be explored in further details in Chapters 9 and 10. Angiogenesis inhibitors are produced by normal tissue cells, but also by tumor cells. The tumor can only progress if it attains mutations which allow for the production of a sufficient level of angiogenesis promoters, such that inhibition is overcome and new blood supply is built. Other, less well defined mechanisms of tumor cell inhibition have also been reported [Guba *et al.* (2001)]. Here we focus on inhibitory effects which the tissue microenvironment can exert on tumor cells, and examine how this influences the relationship between checkpoint gene competence, aging, and the development of cancer.

As before, the model distinguishes between dividing cancer cells, w, and arrested cancer cells, z. These dynamics are captured in the following pair of differential equations which describe the development of the tumor cells over time,

$$\dot{w} = \beta\rho x u\,(1-\eta)\,(1-\alpha) + \frac{rw}{px+1}\left[1-u+u\,(1-\eta)\,(1-\alpha)\,(1-\beta)\right] + \gamma z,$$

$$\dot{z} = \frac{\eta u r w}{px+1} - \gamma z.$$

The first term in the equation for tumor growth represents the production of tumor cells from healthy cells by mutation with a rate $\beta\rho x u(1-\eta)(1-\alpha)$ (see Figure 8.1). In other words, a transformed cell is generated when a healthy cell experiences a genomic alteration which is not repaired, is carcinogenic, and does not result in apoptosis. The parameter r denotes the turnover of the tumor cells. Note the absence of the parameter ϕ which means that there is no constant population size. Instead, the population of tumor cells can grow exponentially. Similarly to healthy cells, tumor cells can become damaged with a rate u. With a probability η the damage results in cell cycle arrest and arrested cells return to a dividing state with a rate γ. With a probability $1-\eta$ damage does not result in cell cycle arrest and repair. With a probability α the damage leads to cell death, for example caused by apoptosis. With a probability $1-\alpha$, the damaged cell survives and attains an additional mutation. This mutation contributes to further cancer progression with a probability β. In addition to these basic cell growth kinetics, the model assumes that tumor cell division is inhibited by the surrounding tissue cells with a rate p. The higher the value of p, the lower the rate of tumor cell division. If the value of p is sufficiently large, the rate of tumor cell division is less than or equal to the tumor death rate; hence, the tumor fails to grow altogether. In the model, only cells which do not repair or are not senescent can contribute to this suppressive activity. This is because repairing or senescent cells are not active. Note that the outcome of the model does not depend on the particular form by which tissue-mediated tumor cell inhibition is described. Alternatives (such as a tissue-mediated increase in the death rate of tumor cells) can also be explored, and give qualitatively identical results (see Chapter 9).

It is important to note that the model only concentrates on describing the initial growth dynamics of the tumor within a tissue environment which varies in checkpoint competence (such as p53+/+, p53+/-, and p53-/- organisms). The checkpoints under consideration (e.g. p53) can induce cell cycle arrest and apoptosis in tumor cells; they can be lost later in the course of progression [Blagosklonny (2002); Kahlem *et al.* (2004)]. Therefore, these initial tumor cells are assumed to have the same level of checkpoint competence as the normal tissue cells.

There are two quantities which are important for the analysis. These are the initial growth rate of the tumor and the mutation rate of the tumor cells. They will be explained in turn.

(i) The initial growth rate of the tumor is made up of two components. These are the production of tumor cells from healthy tissue, and the growth of those tumor cells (clonal expansion). While production of tumor cells by healthy tissue is important for the initiation of tumor cell growth, the growth term becomes dominant as clonal expansion takes off. Because both processes are involved in the initiation of cancer, we consider the time until a tumor reaches an arbitrary threshold size as a measure of how fast a tumor can develop. The longer this time, the slower the development of cancer. In terms of experimental data it correlates with a lower fraction of animals which show a tumor by a defined time point. It is important to note that this growth phase only corresponds to the initial expansion phase which leads to a first stage and detectable tumor. While not included in the model explicitly, this initial growth is not assumed to go on. Instead the tumor is likely to remain at a given small size until further mutations allow the cells to progress and expand further. In the model this is dealt with by stopping the simulation once the tumor has reached this critical threshold size.

(ii) Thus, once a tumor has been initiated and has grown to the threshold size, further progression requires the accumulation of additional carcinogenic mutations. This is influenced by the mutation rate of cells, μ. The higher the mutation rate, the higher the chance that an additional carcinogenic mutation is created. In the model the mutation rate of cells is given by $\mu = \beta ru(1-\eta)(1-\alpha)/(px+1)$.

The following sections will examine how checkpoint competence influences the initial tumor cell growth and the accumulation of further mutations. We will start by examining the basic tumor cell dynamics without taking into account tissue-tumor cell interactions. Then, we will examine how tissue-mediated tumor cell inhibition influences the results. As explained above, the analysis will concentrate on the effect of cell cycle arrest and senescence, captured in the parameter η.

8.5 Checkpoints and basic tumor growth

Here we consider how checkpoint competence influences initial tumor growth and the accumulation of carcinogenic mutations in the absence of tissue-mediated tumor cell inhibition ($p = 0$). That is, the only effect of healthy tissue is to give rise to a transformed cell by mutation. The outcome depends on what can be called the "cost of cell cycle arrest", and the "cost of cell death", and is summarized in Figure 8.3a. These concepts have been explained in detail in Chapter 7. The cost of cell cycle arrest represents the average time cells remain in an arrested state (given by $1/\gamma$). The longer the duration of cell cycle arrest, the more the cell cycle becomes delayed, and the higher the fitness cost for the cells. On the other hand, the cost of cell death represents the fitness reduction of cells which is brought about by the generation of lethal mutants. Lethality can arise from necrotic cell death, but in cancer apoptosis will also play a very important role. The higher the probability that a mutant is lethal (higher value of α), the higher the cost of cell death. The behavior of the model depends on the relative magnitude of these costs.

If the cost of cell cycle arrest is significantly higher than the cost of cell death, a decrease in checkpoint competence correlates with faster initial tumor growth, and with an enhanced ability to accumulate further carcinogenic mutations (Figure 8.3a, 8.4). The reason is that cells in checkpoint deficient organisms avoid senescence, and this leads to faster cell division and a higher mutation rate. As the cost of cell cycle arrest is decreased relative to the cost of cell death (cells return to a dividing state faster and more mutations result in cell death), this picture changes (Figure 8.3a). Now a reduction in checkpoint competence can lead to a slower initial growth of tumor cells. That is, checkpoint deficient organisms are expected to show a reduced incidence of tumors. The lower the cost of cell cycle arrest, and the higher the cost of cell death, the more pronounced this trend. The reason is as follows. The advantage gained from avoiding cell cycle arrest in check-

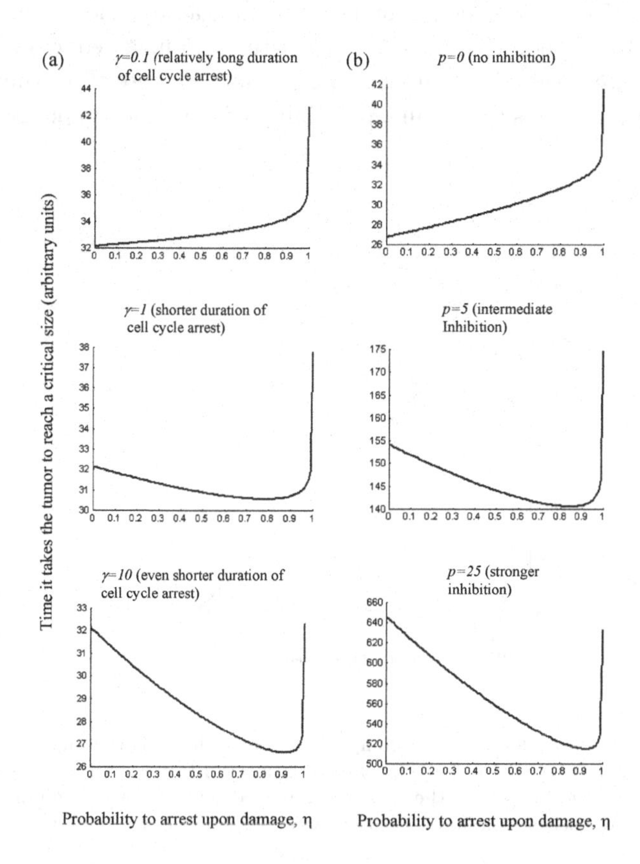

Fig. 8.3 Relationship between checkpoint competence (probability to arrest upon damage, η) and the time it takes for the tumor to reach a critical size (first stage size). We observe similar patterns with two mechanisms. (a) The duration of cell cycle arrest (expressed in the variable γ) is varied. Parameter values were chosen as follows: ρ=2; r=2; u=0.3; α=0.9; $\beta=10^{-7}; d = 0.1; targetsize = 10^{12}$. (b) The amount of tissue induced tumor cell inhibition, p, is varied. Parameter values were chosen as follows: ρ=2; r=2; u=0.3; α=0.5; $\beta=10^{-7}; d = 0.1; targetsize = 10^{12}$.

point deficient organisms is outweighed by the disadvantage of increased cell death in the absence of repair. Overall, this leads to a reduced initial growth rate of tumor cells in the checkpoint deficient scenario. Note that the time it takes for the tumor to reach the threshold size tends towards infinity as $\eta \rightarrow 1$. This is because for η=1, genome surveillance is 100% efficient and no mutants are ever created. This is clearly unrealistic.

While a reduction in checkpoint competence can lead to a slower initial growth rate of the tumor cells in this parameter region, a reduction in checkpoint competence always results in a higher mutation rate of the tumor cells (Figure 8.4), and thus in an enhanced ability of the tumor to progress.

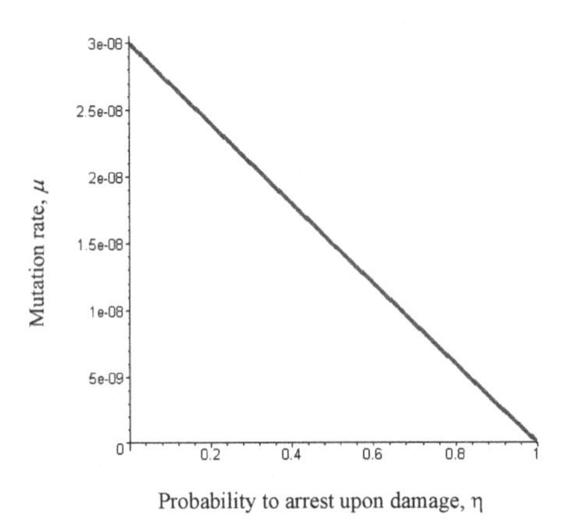

Fig. 8.4 Relationship between checkpoint competence (probability to arrest upon damage, η) and the mutation rate of cells. The same relationship is observed irrespective of the duration of cell cycle arrest or the rate of tissue-mediated tumor cell inhibition. Parameter values were chosen as follows: ρ=*2*; *r*=*2*; *u*=*0.3*; α=*0.9*; β=*10*$^{-7}$; $d = 0.1$.

Therefore, if the cost of cell death is high compared to the cost of cell cycle arrest, the model gives rise to the counter-intuitive observation that a reduction in checkpoint competence can lead to the generation of fewer first-stage tumors during a defined time frame; however, the tumors which do arise are expected to accumulate further carcinogenic mutations faster and therefore have the potential to progress faster. It remains to be determined how relevant this result is. While these conditions could hold for checkpoint genes which are solely responsible for DNA repair, they might not hold for checkpoints which are responsible for apoptosis, or for both senescence and apoptosis (such as p53). If a reduction in the checkpoint leads to reduced apoptosis, cell death in not likely to be the dominant fitness cost.

8.6 Tumor growth and the microenvironment

Now we examine how tissue-mediated tumor cell inhibition influences the effect of checkpoint competence on the initial growth rate of the tumor and the mutation rate of tumor cells. The dynamics depend on the degree with which healthy tissue exerts inhibitory activity on the tumor cells, p. This is shown in Figure 8.3b, and also summarized by time-series simulations in Figure 8.5.

First, assume that the tissue-mediated inhibitory activity is relatively small and lies below a threshold (small value of p). Now the behavior of the model is identical to the one described in the last section. We concentrate on the parameter region in which a reduction in checkpoint competence leads to an increased initial growth rate of the tumor cells, and to an increased mutation rate (shown again in Figure 8.3b for reference). In other words, senescence represents an important cost for the cells. As explained above, reduced checkpoint competence allows cancer cells to proceed through the cell cycle without delay following genomic alterations (Figure 8.5a). Again, at values of η close to 1, the time it takes the tumor to reach the threshold size becomes rapidly longer and goes to infinity for η=*1* because genomic surveillance is 100% efficient (Figure 8.3b).

As the rate of tumor cell inhibition, p, is increased, these patterns change (Figure 8.3b). Now, a reduction of checkpoint competence results in a slower growth rate of the tumor (summarized by a time-series graph in Figure 8.5b). The higher the checkpoint competence, the faster the initial growth rate of the tumor. As the rate of tissue-mediated inhibition of cancer cells is increased further, the more pronounced this relationship becomes (Figure 8.3b). This is the same counter-intuitive result observed above, and is explained as follows in the current context. A checkpoint has two opposite effects on cancer cells in the model. On the one hand, it sends tumor cells into cell cycle arrest and this slows down tumor growth. On the other hand, it reduces the inhibitory effect exerted by tissue cells, and this enhances tumor growth. Tissue mediated inhibitory effects are reduced because the tissue cells frequently enter cell cycle arrest during which they do not function at normal levels. If tissue-mediated inhibition plays a sufficiently important role in the dynamics of tumor initiation (high values of p), the advantage gained by reduced tissue-mediated inhibition in the presence of a checkpoint outweighs the cost the cancer cells have to pay as a result of checkpoint competence. As before, we observe that as the value of η approaches 1, the time it takes for the tumor to grow to the threshold size

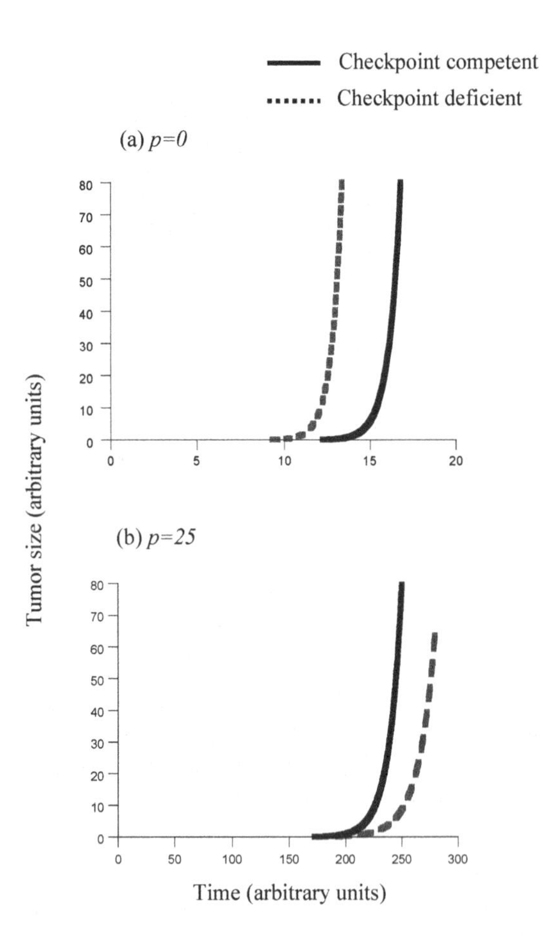

Fig. 8.5 Time series depicting the average initial growth of the first stage tumor in checkpoint competent organisms (solid line) and checkpoint deficient organisms (dashed line). (a) Rate of tissue-mediated tumor cell inhibition is zero. (b) Rate of tissue-mediated tumor cell inhibition is relatively high. Parameter values were chosen as follows: $\rho=2$; $r=2$ $\gamma=0.1$; $u=0.3$; $\alpha=0.5$; $\beta=10^{-7}$; $d=0.1$. Checkpoint competent organisms have $\eta=0.9$, and checkpoint deficient organisms have $\eta=0.1$.

goes towards infinity because genomic surveillance prevents the generation of any tumor cells (Figure 8.3b). In contrast to the rate of tumor growth, an increase in the level of tissue-mediated tumor cell inhibition does not change the relationship between checkpoint competence and the mutation rate of the cells (ability of the tumor to progress). Regardless of the value of p, a reduction in checkpoint competence is predicted to result in a faster rate of mutation (Figure 8.4). Therefore while reduced checkpoint competence is expected to result in the establishment of fewer tumors within a given time frame, the tumors which do develop are predicted to progress faster. This is the same outcome as observed in the last section, but is brought about by a different mechanism.

The only case where these relationships do not hold is at very high rates of DNA damage ($u \rightarrow 1$). In this case, most tissue cells will be non-functional or dead and tumor cell inhibition is not a significant factor anymore. This parameter region is, however, biologically unrealistic because such a scenario would correspond to death of the organism as a result of senescence or tissue destruction.

While not considered explicitly, the same considerations should apply to checkpoints which induce apoptosis. Higher levels of apoptosis can reduce the number of tissue cells, and this can lead to a compromised ability of the tissue to inhibit tumor cells. If tissue-mediated inhibition of tumor cells is a sufficiently significant component in the dynamics of tumor initiation, then apoptosis deficiency can lead to reduced tumor incidence, but in faster progression of the tumors which do develop. Whether this argument holds depends on how exactly apoptosis influences tissue size and the process of aging [Campisi (2003a); Camplejohn *et al.* (2003); Gilhar *et al.* (2004); Jansen-Durr (2002)].

8.7 Theory and data

The following relationships between checkpoint competence, senescence, and the development of cancer were found.

(1) The higher the checkpoint competence, the lower the rate of cancer incidence and progression, but the earlier the onset of aging (level of senescence is higher in the face of DNA damage). Lower checkpoint competence prevents an early onset of aging, but promotes cancers. This outcome is brought about by the following conditions. First, the advantage derived from avoiding cell cycle arrest upon damage must be higher than the cost derived from cell death in the absence of repair. In addition it requires that tissue cells exert no (or only little) inhibitory activity on tumor cells. These arguments might only apply to checkpoints which induce cell cycle arrest, and not to checkpoints which induce apoptosis.

(2) A reduction in checkpoint competence can result both in fewer aging symptoms and in fewer cancers; however, the cancers which do become established are characterized by accelerated progression. This is promoted by the following conditions (for a schematic explanation see Figure 8.6). First, it may occur because the cost derived from increased cell death in the absence of repair is higher than the benefit derived from avoiding cell cycle arrest upon damage. Whether this argument applies depends on the details of the checkpoint mediated activity, and requires that cell death (and thus apoptosis) occurs efficiently. However, a separate mechanism can lead to the same observation. It involves tissue-mediated tumor cell inhibition as a significant factor in the dynamics of tumor initiation. Reduced checkpoint competence prevents tissue aging and preserves the inhibitory function. This in turn leads to the development of fewer tumors. In contrast, higher checkpoint competence promotes tissue aging and compromises inhibitory function. In this parameter region, the advantage which the tumor cells gain from impaired inhibition outweighs the cost derived from the disruption of the tumor cell cycle; this leads to the generation of more cancers. This effect of tissue-mediated tumor cell inhibition is summarized by computer simulations in Figure 8.5. Since this mechanism requires that checkpoint-induced aging plays a dominant role in the dynamics of tissue cells, it is promoted by relatively high levels of DNA damage which can potentially trigger the checkpoints. It is unlikely to work if DNA damage is a relatively rare event. This mechanism may also be relevant to the induction of apoptosis if apoptosis contributes to tissue aging and a decline of tissue function.

In the following, these notions will be discussed in the context of mice which have varying competence to mount p53 responses upon genomic damage.

A recent study has demonstrated that p53 might be an important factor which contributes to senescence and aging [Donehower (2002); Tyner *et al.* (2002)], and this has sparked much discussion [Kirkwood (2002); Sharpless and DePinho (2002)]. Tyner et al. developed a genetically altered mouse that can express a truncated form of p53 which augments wild-type p53 activity. Survival of these super-p53 mice was compared to wild-type (p53+/+) animals as well as p53 deficient animals (p53-/+, p53-/-). The experiments demonstrated that p53+/+ mice showed best survival, followed by the super-p53 mice. The p53 deficient animals were characterized

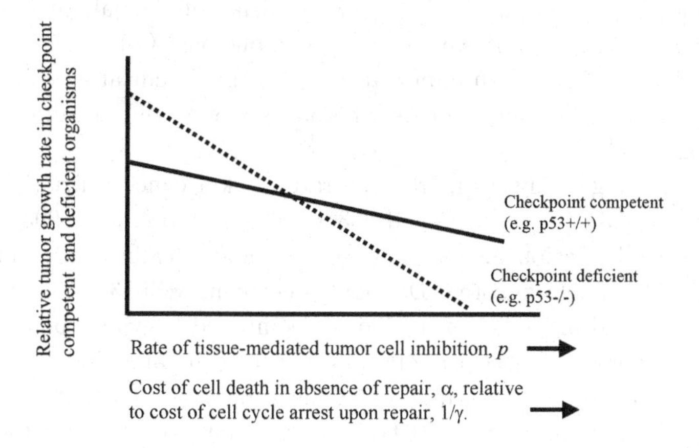

Fig. 8.6 Schematic summary of the counter-intuitive mathematical result regarding checkpoint competence and tumor incidence. This is not based on numerical simulation, but is a graphical summary to aid understanding.

by an even lower survival rate because of the early development of certain cancers. It turned out that the super-p53 animals had reduced survival compared to wild-type mice because they experienced an accelerated onset of aging. On the other hand, tumor incidence in super-p53 mice was greatly reduced. These experiments support the notion that p53 activity represents a tradeoff between preventing senescence and preventing the development of cancer. This corresponds to one of the parameter regions observed in the model, and the conditions of the experiments are consistent with this parameter region. Two aspects are likely to be responsible. First, the study looked for the spontaneous development of tumors and did not induce them with carcinogenic agents. This means that especially at relatively early ages of the mice, DNA damage and the induction of checkpoints are rare events in healthy tissue cells. Consequently, the amount of tissue-mediated tumor cell inhibition is not likely to differ significantly between p53 competent and p53 deficient mice during this phase. The only effect of p53 deficiency is to allow the tumor cells to progress faster through the cell cycle in the face of oncogenic mutations. Therefore, p53 deficient mice are expected to show an elevated onset of tumors during early life. In addition, it is possible that the cancers which developed in the p53 deficient mice are not significantly inhibited by tissue cells. The amount of tissue-mediated inhibition of tumor cells can vary between different tissues. In this context, it is interesting to note that the cancers which developed in p53 deficient animals in this study were specific cases, mostly sarcomas and lymphomas. Other cancers in which the deletion of p53 is an important step did not occur at elevated levels. In fact, lung cancers were observed in p53 competent but not in p53 deficient animals.

Another interesting study compared the rate of skin cancer initiation and progression in p53+/+, p53+/-, and p53-/- mice [Kemp *et al.* (1993)]. The rate of cancer initiation and progression was statistically similar in p53+/+ and p53+/- heterozygotes. Double knockout mice (p53-/-), however, showed a reduced incidence of papillomas compared to wild-type animals. On the other hand, the tumors which were generated in p53-/- mice were characterized by more rapid malignant progression compared to p53+/+ animals. This is the second type of behavior predicted by the model. The reason that this behavior is observed in the Study by Kemp et al. could be as follows. First, the development of skin cancer is known to depend strongly on angiogenesis, and therefore inhibition of angiogenesis could ensure that tissue-mediated tumor cell inhibition is a significant force in the process of cancer initiation and progression. Moreover, in contrast to the study by Tyner et al., mice were treated with carcinogenic agents in order to induce tumors. This means that mice experience elevated levels of DNA damage and frequent induction of p53. This can result in elevated levels of tissue senescence in p53 competent animals, and this could result in reduced ability of tissue cells to display anti-tumor activity. Hence, tumors develop more often in p53 competent, compared to p53 deficient mice. According to the model, another explanation for this outcome could be that in the absence of repair, the cost derived from the production of lethal mutants outweighs the benefit derived from avoiding repair and cell cycle arrest. It is, however, not clear whether this is a likely explanation. Because p53 is also involved in the induction of apoptosis upon genomic damage or oncogenic mutations, it can be argued that p53 deficient organisms show reduced levels of apoptosis and cell death. Therefore, lethality in the absence of repair might not be a dominant factor.

This work also has implications for understanding the pattern of cancer incidence in patients which have a familial genetic defect in checkpoint genes. An interesting example is Li-Fraumeni syndrome which is characterized by lack of functional p53 in every cell of the body [Evans and Lozano (1997)]. Interestingly, patients develop only certain types of cancers, most importantly sarcomas. Although other types of cancers, such as colon cancer, also involve p53 inactivation as a crucial event in progression, they do not occur at elevated rates in Li-Fraumeni patients. According to the modeling results presented here, the exact details involved in the process of carcinogenesis can determine whether reduced p53 activity (and reduced checkpoint competence in general) leads to a higher incidence of cancers or not.

Chapter 9

Basic models of tumor inhibition and promotion

The development of cancer is regulated on many levels. So far, we have concentrated on the development of the cancerous phenotype itself. That is we investigated the processes which lead to the generation of a malignant cell and examined conditions under which genetically unstable cells can emerge. However, even if cancer cells have been generated and can in principle evolve to accumulate more mutations, these cancer cells might not be able to grow beyond a very small size. The reason is that the body is characterized by specific defenses which try to prevent the growth and pathogenicity of selfish transformed cells once they have been generated. In particular, the microenvironment in which the cancer emerges is thought to play a pivotal role in deciding whether the cancer will succeed at growing to high levels or not [Tlsty (2001); Tlsty and Hein (2001)]. Indeed, the development of cancer may require a conspiracy between tumor cells and their microenvironment [Hsu *et al.* (2002)].

One of the most important players in this respect is the blood supply which provides cancer cells with oxygen, the necessary nutrients, and factors required for replication and survival. A given tissue or organ must have a sufficient blood supply in order to function. No extra blood supply is available though, which will hinder any potential abnormal growth. Cancer cells have to induce the generation of new blood supply in order to sustain their growth. This process is called angiogenesis (as explained already in Chapter 1). Research on the role of angiogenesis for cancer progression has been pioneered by Judah Folkman in the 1960s and 70s [Folkman (1971)], and work from his laboratory has been dominating the literature up to now (e.g. [Folkman (1995a); Folkman (2002)]). In early experiments, Folkman and colleagues placed a small number of rabbit melanoma cells on the surface of the rabbit thyroid gland. They observed that the tumor cells initially grew but subsequently stopped growing once they reached a relatively small size comparable to that of a pea. The reason is that the tumor cells run out of blood supply.

It is now clear that growth to larger sizes requires the emergence of so–called *angiogenic tumor cells*. The ability of the cancer to grow depends on the balance between so–called angiogenesis inhibitors, and angiogenesis promoters. Examples of inhibitors are thrombospondin, tumstatin, canstatin, endostatin, angiostatin and interferons. Examples of promoters are growth factors such as FGF, VEGF, IL-8, and PDGF. Normal tissue produces mostly angiogenesis inhibitors. So do cancer cells. This serves as a preventative measure against abnormal growth. Angiogenic cancer cells, on the other hand, have mutations which allow the balance between inhibitors and promoters to be shifted away from inhibition, and towards promotion. This is done by activating the production of angiogenesis promoters, or by inactivating genes which encode inhibitors. Once such angiogenic cells have evolved, it is possible for the cancer to recruit new blood vessels and hence to grow to larger sizes. Folkman's research has also given rise to exciting new avenues of therapies [Hahnfeldt *et al.* (1999b)]: Administration of angiogenesis inhibitors can destroy blood supply and result in remission of cancers. While encouraging results have been obtained in laboratory animals, our understanding is far less complete in the context of human pathologies.

This chapter reviews mathematical models which have examined the dynamics of angiogenesis-dependent tumor growth. The absence of blood supply can affect tumor cells in two basic ways [Folkman (2002)]. On the one hand, it can increase the death rate of tumor cells. In the absence of blood supply, apoptosis can be triggered as a result of hypoxia. On the other hand, the absence of blood supply can prevent cell division and growth. In this case the cells are dormant; that is, they do not divide or die. Both types of scenarios have been modeled and will be discussed. The models have similar properties, and we will discuss the requirements for the evolution of angiogenic cell lines and for the transition from a small and non-pathogenic tumor to a tumor with malignant potential. We will then take one of the models and incorporate the spread of the tumor across space into the equations. We will discuss how the dynamics between tumor promotion and inhibition influence more advanced tumor growth within a tissue. Finally, we will discuss clinical implications of the modeling results.

9.1 Model 1: Angiogenesis inhibition induces cell death

We describe and analyze a model for the evolution of angiogenic tumor cell lines [Wodarz and Krakauer (2001)]. The model consists of three basic variables (Figure 9.1).

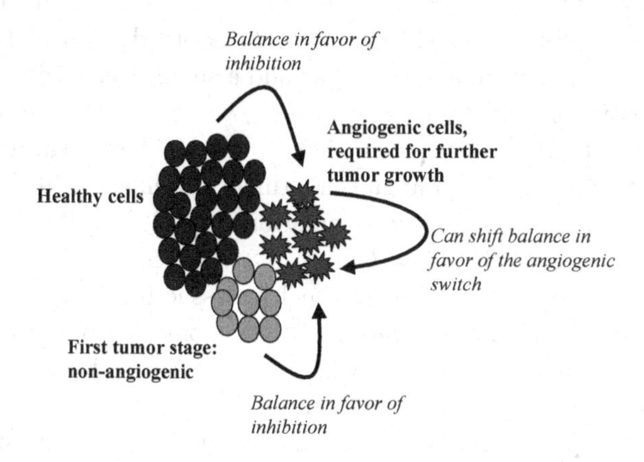

Fig. 9.1 Schematic diagram illustrating the central assumptions underlying the mathematical model.

Healthy host tissue, x_0; a first transformed cell line, x_1, which is non-angiogenic and cannot grow above a given threshold size; an angiogenic tumor cell line which has the potential to progress, x_2. It is thought that the formation of new blood vessels depends on a balance of angiogenesis inhibitors and promoters. If the balance is in favor of the inhibitors, new blood vessels are not formed. On the other hand, if it is in favor of the promoters, angiogenesis can proceed. Hence, the model assumes that healthy tissue, x_0, and stage one tumor cells, x_1, produce a ratio of inhibitors and promoters that is in favor of angiogenesis inhibition. On the other hand, it is assumed that angiogenic tumor cell lines have the ability to shift the balance in favor of angiogenesis promotion. We first consider progression from the wildtype cells to a first transformed cell line. The basic model is given by the following pair of differential equations,

$$\dot{x}_0 = r_0 x_0 \left(1 - \frac{x_0}{k_0}\right)(1 - \mu_0) - d_0 x_0,$$

$$\dot{x}_1 = \mu_0 r_0 x_0 \left(1 - \frac{x_0}{k_0}\right) + r_1 x_1 \left(1 - \frac{x_1}{k_1}\right)(1 - \mu_1) - d_1 x_1.$$

Healthy cells are assumed to replicate at a density dependent rate $r_0 x_0(1 - x_0/k_0)$. The value of k_0 represents the maximum size this population of cells can achieve, or the carrying capacity. The cells die at a rate $d_0 x_0$. We assume that the rate of mutation is proportional to the rate of replication of the cells, and is thus given by $\mu_w r_0 x_0(1 - x_0/k_0)$. The mutations give rise to the first stage of tumor progression, x_1, i.e. to a tumor cell line that is not angiogenic. This cell line will depend on the blood supply of the healthy tissue and will not be able to grow beyond a small size. These cells replicate at a density dependent rate $r_1 x_1(1 - x_1/k_1)$, where the carrying capacity k_1 is assumed to be relatively small ($k_1 << k_0$). They die at a rate $d_1 x_1$, and mutate to give rise to an angiogenic tumor cell line, x_2, at a rate $\mu_x r_1 x_1(1 - x_1/k_1)$. In the model, the population of healthy cells attains a homeostatic setpoint given by $x_0^* = k_0(r_0 - d_0)/r_0$. The mutation rate μ_0 can be assumed to be very small, since healthy tissue has intact repair mechanisms that ensure faithful replication of the genome. Once mutation gives rise to the first tumor cell line, it will grow to its small homeostatic set point level defined by $x_1^* = k_1(r_1 - d_1)/r_1$.

The wildtype cell population and the small population of first stage tumors are assumed to reach constant levels in a relatively short time. In other words, they reach an equilibrium abundance. Further tumor growth requires the emergence of the angiogenic cell line, x_2. In the following we investigate the conditions required for angiogenic tumor cell lines to evolve assuming a constant background abundance of x_0 and x_1.

The angiogenic cell line replicates at a density dependent rate $r_2 x_2(1 - x_2/k_2)$. As these cells can potentially influence the balance of inhibitors and promoters in favor of promoters, we have to take these dynamics into account. The death rate of these cells is determined by two components. The angiogenic tumor cells are characterized by a composite background death rate $d_2 x_2$, as with x_0 and x_1. In addition, the model assumes that the death rate can be increased if the balance between angiogenesis inhibition and promotion is in favor of inhibition. Hence, this death rate is expressed as $(p_0 x_0 + p_1 x_1 + p_2 x_2)/(q x_2 + 1)$. Thus all three cell types lead to the inhibition of angiogenesis, whereas inhibition of angiogenesis can only be overcome by cell line x_2.

As we have assumed that x_0 and x_1 are at equilibrium, we start our analysis by ignoring mutation and simply looking at the dynamics of the angiogenic cell line, x_2. These dynamics are described by the equation,

$$\dot{x}_2 = r_2 x_2 \left(1 - \frac{x_2}{k_2}\right) - d_2 x_2 - \frac{x_2 \left(p_0 x_0^* + p_1 x_1^* + p_2 x_2\right)}{q x_2 + 1},$$

where x_0^* and x_1^* are defined above. Two outcomes are possible. (i) The cell line x_2 cannot invade, resulting in equilibrium E0 where $x_2^{(0)} = 0$. (ii) The cell line x_2 can invade and converges to equilibrium E1 described by

$$x_2^{(1)} = \frac{-Q + \sqrt{Q^2 - 4 r_2 q_2 \left[k \left(d_2 - r_2\right) + k p_I x_I\right]}}{2 r_2 q_2},$$

where $Q = k q_2 \left(d_2 - r_2\right) + r_2 + k p_2$ and subscript I refers to the inhibitory cell lines: $p_I x_I = p_0 x_0^* + p_1 x_1^*$.

In the following we examine the stability properties of these two equilibria which are summarized in Figure 9.2. If $p_I x_I < r_2 - d_2$, then the equilibrium describing the extinction of the angiogenic tumor, $x_2^{(0)}$, is not stable. The equilibrium describing the invasion of the angiogenic tumor cell line, $x_2^{(1)}$, is stable. In other words, if the above condition is fulfilled, then the degree of angiogenesis inhibition is too weak, and the angiogenic tumor cell line can emerge, marking progression of the disease.

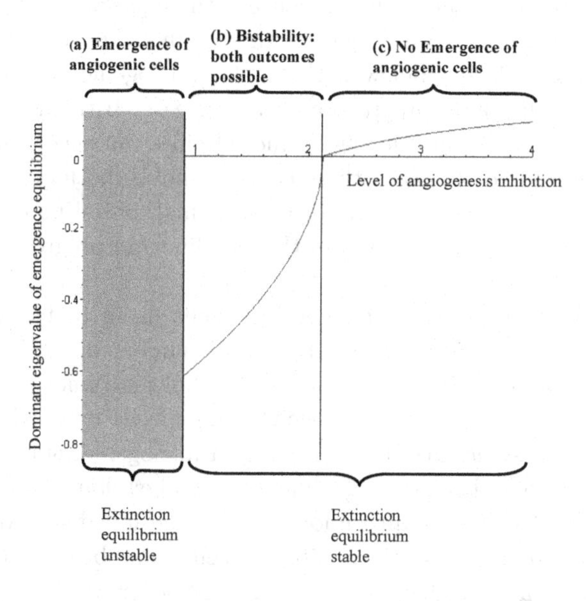

Fig. 9.2 Graph showing the stability properties and the outcome of the model. Parameters were chosen as follows: $r_2 = 1; k_2 = 1; d_2 = 0.1; p_2 = 1; q = 10$.

On the other hand, if $p_I x_I > r_2 - d_2$, the degree of inhibition is stronger and the situation is more complicated (Figure 9.2). The equilibrium describing the extinction of the angiogenic cell population, $x_2^{(0)}$, becomes stable. However, equilibrium $x_2^{(1)}$, describing the emergence of angiogenic tumor cells, may or may not be stable (Figure 9.2).

(1) If the degree of angiogenesis inhibition lies above a certain threshold, equilibrium $x_2^{(1)}$ is unstable and the angiogenic cell line cannot invade. It was not possible to define this threshold in a meaningful way.
(2) If the degree of angiogenesis inhibition lies below this threshold, equilibrium $x_2^{(1)}$ remains stable. Now, both the extinction and the emergence equilibria are stable (Figure 9.2). This means that two outcomes are possible and that the outcome depends on the initial conditions. Either the angiogenic cell line fails to emerge, or the angiogenic cell line does emerge, resulting in tumor progression. As shown in Figure 9.3, a low initial abundance of angiogenic tumor cells results in failure of growth. On the other hand, a high initial number of angiogenic tumor cells results in growth of the tumor and progression (Figure 9.3).

To summarize, the model shows the existence of three parameter regions (Figure 9.2). If the degree of angiogenesis inhibition by healthy tissue and stage one tumor cells lies below a threshold, angiogenic tumor cell lines always invade resulting in progression of the disease. If the degree of inhibition lies above a threshold, the angiogenic cell lines can never emerge and pathology is prevented. Between these two thresholds, both outcomes are possible depending on the initial conditions. A high initial number of angiogenic tumor cells results in growth of this cell line and progression of the disease.

What does the initial number of angiogenic cells mean in biological terms? The dependence of growth on the initial number of angiogenic tumor cells presents an effective barrier against pathologic tumor growth. Given that a small number of non-angiogenic tumor cells exists, it will be difficult to create a sufficiently large number of angiogenic mutants to overcome the blood supply barrier. This difficulty could explain why, upon autopsy, people tend to show small tumors which have failed to grow to larger sizes. The initial number of the angiogenic cells could be determined by the mutation rate μ_1, which gives rise to the angiogenic cells. If the mutation rate is high, the initial number of angiogenic cells will be high. On the other hand, if the mutation rate is low, the initial number of the

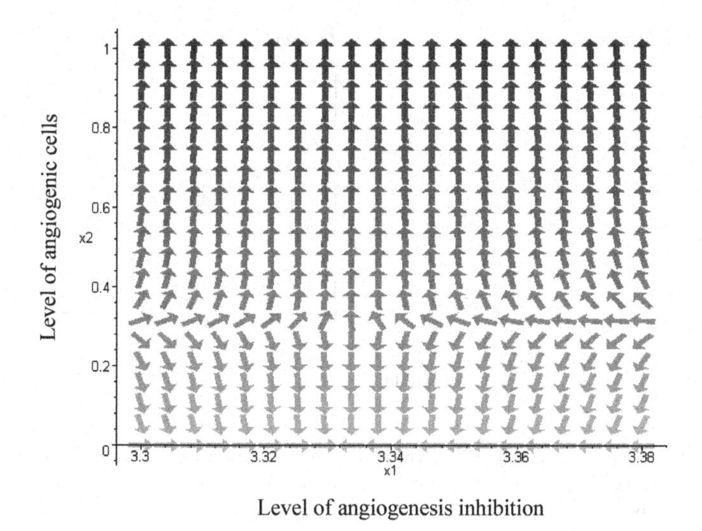

Fig. 9.3 Direction field plot showing how the outcome of the model can depend on the initial conditions. Parameters were chosen as follows: $r_2 = 1; k_2 = 100; d_2 = 0.1; p_2 = 1; q = 10$. For the purpose of simplicity the populations of non-angiogenic cells were summarized in a single variable and assumed to converge towards a stable setpoint (characterized by the parameters *r=0.15; d=0.1; k=10*).

angiogenic cells will be low. Hence, in the parameter region where the outcome of the dynamics depends on the initial conditions, a high mutation rate promotes the emergence and growth of angiogenic tumor cells (Figure 9.4). If a high mutation rate by tumor cells defines genetic instability, then it is possible that genetic instability might be required for the invasion of angiogenic tumor cells.

9.2 Model 2: Angiogenesis inhibition prevents tumor cell division

We consider a basic mathematical model which describes the growth of a cancer cell population, assuming that the amount of blood supply does not influence cell death, but the rate of cell division [Wodarz and Iwasa (2004)]. This model will also be used to consider the effect of diffusion of cells and soluble molecules across space; this is done in the next sections. Therefore, the model will take into account explicitly the dynamics of promoters and inhibitors. This is in contrast to the last section where for the purpose

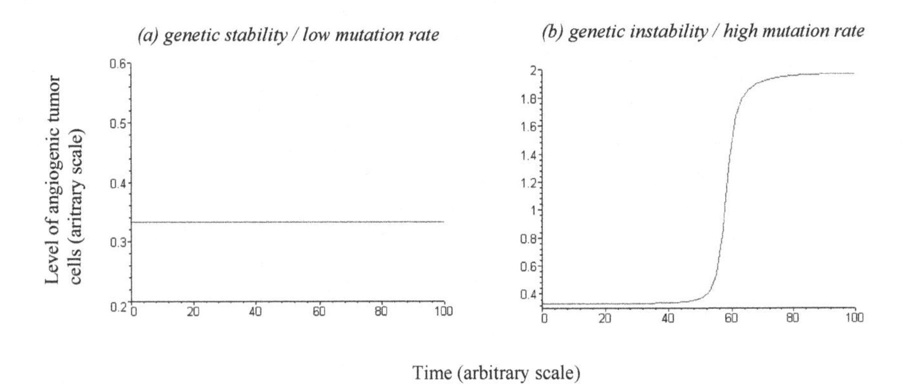

Fig. 9.4 Genetic instability and the emergence of angiogenic cell lines. (a) If the mutation rate is low (genetic stability), the initial number of angiogenic cells created is low. Consequently they cannot emerge. (b) On the other hand, if the mutation rate is high (genetic instability), a higher initial number of angiogenic cells is created. Hence, they emerge and become established. Parameters were chosen as follows: $r_0 = 0.11; k_0 = 10; \mu_0 = 0.001; d_0 = 0.1; r_1 = 0.12; k_1 = 2; d_2 = 0.1; r_2 = 2.5; k_2 = 2; d_2 = 0.1; p_0 = 2; p_1 = 2; p_2 = 2; q = 10$; for (a) $\mu_1 = 0.001$; For (b) $\mu_1 = 0.01; \mu_2 = \mu_1$.

of simplicity inhibitors and promoters were assumed to be proportional to the number of cells which secrete them. The new model includes three variables: the population of cancer cells, C; promoters, P; and inhibitors, I. It is assumed that both promoters and inhibitors can be produced by cancer cells. In addition, inhibitors may be produced by healthy tissue. The model is given by the following set of differential equations which describe cancer growth as a function of time,

$$\dot{C} = \left(\frac{rC}{\epsilon C + 1}\right)\left(\frac{P}{I+1}\right) - \delta C, \tag{9.1}$$

$$\dot{P} = a_P C - b_P P, \tag{9.2}$$

$$\dot{I} = \xi + a_I C - b_I I. \tag{9.3}$$

The population of cancer cells grows with a rate r. Growth is assumed to be density dependent and saturates if the population of cancer cells becomes large (expressed in the parameter ϵ). In addition, the growth rate of the cancer cells depends on the balance between promoters and inhibitors, expressed as $P/(I+1)$. The higher the level of promoters relative to inhibitors, the faster the growth rate of the cancer cell population. If the level of promoters is zero, or the balance between promoters and inhibitors in heavily in favor of inhibitors, the cancer cells cannot grow and remain dormant [O'Reilly *et al.* (1997); O'Reilly *et al.* (1996); Ramanujan *et al.* (2000)]. Cancer cells are assumed to die at a rate δ. Promoters are produced by cancer cells at a rate a_p and decay at a rate b_p. Inhibitors are produced by cancer cells at a rate a_I and decay at a rate b_I. In addition, the model allows for production of inhibitors by normal tissue at a rate ξ.

9.2.1 *Linear stability analysis of the ODEs*

Let us simplify system (9.1-9.3) by using a quasistationary approach, that is, we will assume that the level of promoters adjusts instantaneously to its steady-state value ($P = Ca_P/b_P$). It is convenient to denote

$$W = \frac{ra_P}{\delta b_P}, \quad \gamma = \frac{a_I}{b_I}.$$

Now we have a two-dimensional system,

$$\dot{C} = \delta C\left(\frac{WC}{(1+\epsilon C)(1+I)} - 1\right), \tag{9.4}$$

$$\dot{I} = b_I(\gamma C - I). \tag{9.5}$$

There can be up to three fixed points in this system,

$$(C, I) = (0, 0), \quad \text{and} \quad (C, I) = (\hat{C}_\pm, \hat{I}_\pm),$$

where $\hat{I}_\pm = \gamma \hat{C}_\pm$, and

$$\hat{C}_\pm = \frac{-(\gamma + \epsilon - W) \pm \sqrt{(\gamma + \epsilon - W)^2 - 4\epsilon\gamma}}{2\epsilon\gamma}.$$

It is obvious that if $\gamma + \epsilon - W < 0$, and $(\gamma + \epsilon - W)^2 - 4\epsilon\gamma > 0$, then there are exactly three positive equilibria in the system. If either of these conditions is violated, the $(0, 0)$ solution is the only (biologically meaningful) stable point.

Stability analysis can be performed by the usual methods. It shows that for the $(0, 0)$ equilibrium, the Jacobian is

$$\begin{pmatrix} -\delta & 0 \\ b_I\gamma & -b_I \end{pmatrix},$$

that is, this equilibrium is always stable. For the points $(C_\pm, I_\pm)$, we get the following Jacobian,

$$\begin{pmatrix} \frac{-\delta(\epsilon-\gamma-W\pm\Gamma)}{2W} & \frac{\delta(\epsilon-\gamma-W-\Gamma)}{2\gamma W} \\ b_I\gamma & -b_I \end{pmatrix},$$

where we denote for convenience, $\Gamma \equiv \sqrt{(\epsilon+\gamma-W)^2-4\epsilon\gamma}$. It is easy to show that the eigenvalues of this matrix for the solution $(\hat{C}_-, \hat{I}_-)$ are given by

$$\frac{1}{4W}\left(-Y_- \pm \sqrt{Y_-^2 + 16b_I\delta W\Gamma}\right),$$

and for the solution $(\hat{C}_+, \hat{I}_+)$ we have eigenvalues

$$\frac{1}{4W}\left(-Y_+ \pm \sqrt{Y_+^2 - 16b_I\delta W\Gamma}\right),$$

where $Y\pm \equiv 2b_IW + \delta(\epsilon-\gamma-W\pm\Gamma)$. We can see that solution $(\hat{C}_-, \hat{I}_-)$ is always unstable and we will not consider it any longer. Solution $(\hat{C}_+, \hat{I}_+)$, which we call for simplicity $(\hat{C}, \hat{I})$ from now on, is stable as long as

$$Y_+ > 0. \tag{9.6}$$

9.2.2 *Conclusions from the linear analysis*

As we can see this model has very similar properties compared to the last one, and they are summarized as follows. There are two outcomes. *(i)* The cancer cells cannot grow and consequently go extinct. That is, $C = 0$, $P = 0$ and $I = \xi/b_I$. The cancer goes extinct in the model because we only consider cells which require the presence of promoters for division. If the level of promoters is not sufficient, the rate of cell death is larger than the rate of cell division. In reality, however, it is possible that a small population of non-angiogenic tumor cells survives. This was modeled in more detail in the previous section. Here, we omit this for simplicity. *(ii)* The population of cancer cells grows to significant levels, that is, $C = \hat{C}$.

How do the parameter values influence the outcome of cancer growth? The cancer extinction outcome is always stable. The reason is as follows. The cancer cells require promoters to grow. The promoters, however, are produced by the cancer cells themselves. If we start with a relatively low initial number of cancer cells, this small population cannot produce enough promoters to overcome the presence of inhibitors. Consequently, the cancer fails to grow and goes extinct. This outcome is always a possibility, regardless of the parameter values. Significant cancer growth can be observed if the intrinsic growth rate, r, lies above a threshold relative to the death rate of the cells, δ, and degree of tumor cell inhibition (a_p and b_p relative to a_I and b_I). The exact condition is given by (9.6). In this case, the outcome is either failure of cancer growth, or successful growth to large numbers. Which outcome is achieved depends on the initial conditions. Successful growth is only observed if the initial number of cancer cells lies above a threshold. Then, enough promoters are initially produced to overcome inhibition. This is the same result as presented in the previous section; in biological terms this may mean that mutant cells which produce promoters must be generated frequently (e.g. by mutator phenotypes) in order to initiate tumor growth to higher levels [Wodarz and Krakauer (2001)].

9.3 Spread of tumors across space

In this section, we introduce space into the above described model. We consider a one-dimensional space along which tumor cells can migrate. The model is formulated as a set of partial differential equations and is written as follows,

$$\frac{\partial C}{\partial T} = \left(\frac{rC}{\epsilon C + 1}\right)\left(\frac{P}{I+1}\right) - \delta C + D_c \frac{\partial C^2}{\partial x^2}, \tag{9.7}$$

$$\frac{\partial P}{\partial T} = a_P C - b_P P, \tag{9.8}$$

$$\frac{\partial I}{\partial T} = a_I C - b_I I + D_I \frac{\partial I^2}{\partial x^2}, \quad 0 \leq x \leq L. \tag{9.9}$$

The model assumes that tumor cells can migrate, and this is described by the diffusion coefficient D_c. Inhibitors can also diffuse across space, and this is described by the diffusion coefficient D_I. It is generally thought that inhibitors act over a longer range, while promoters act locally [Folkman (2002); Ramanujan *et al.* (2000)]. Therefore, we make the extreme assumption that promoters do not diffuse. For simplicity we assume that inhibitors are only produced by cancer cells and ignore the production by normal tissue (that is, $\xi = 0$). This simplification is justified because this model concentrates on the tumor dynamics, and numerical simulations show that the results considered here are not altered by this simplification. As mentioned above, the model considers tumor spread across space. It is important to point out that we do not consider long-range metastatic spread. Instead, we consider local spread of a tumor within a tissue, such as the breast, liver, brain, or esophagus.

These equations must be equipped with appropriate initial and boundary conditions. In the simulations we used the following (Neumann) boundary conditions:

$$\left.\frac{\partial C}{\partial x}\right|_{x=0} = \left.\frac{\partial I}{\partial x}\right|_{x=0} = \left.\frac{\partial P}{\partial x}\right|_{x=0} = \left.\frac{\partial C}{\partial x}\right|_{x=L} = \left.\frac{\partial I}{\partial x}\right|_{x=L} = \left.\frac{\partial P}{\partial x}\right|_{x=L} = 0.$$

The dependence of the results on the initial conditions is discussed below.

Here we investigate the process of tumor growth and progression in relation to the degree of inhibition and promotion. First we will present a mathematical analysis and then biological insights and results of simulations.

9.3.1 *Turing stability analysis*

Again, we are going to assume that promoters adjust instantaneously to their equilibrium level. By replacing P with C defined by $P = \frac{a_P}{b_p} C$, we can rewrite equation (9.7) as

$$\frac{\partial C}{\partial t} = \left(\frac{Cra_p}{b_p(1+\epsilon C)(1+I)} - \delta \right) C + D_y \frac{\partial^2 C}{\partial x^2}. \tag{9.10}$$

This equation together with equation (9.9) gives a Turing model.

Let us go back to the system of ODEs, (9.4-9.5), and assume that solution $(\hat{C}, \hat{I})$ is a stable equilibrium. Of course, this solution also satisfies the system of PDEs, (9.10,9.9). Let us consider a wave-like deviation from this spatially uniform solution:

$$C(x,t) = \hat{C} + A\cos(\omega x)e^{\lambda t},$$

$$I(x,t) = \hat{I} + B\cos(\omega x)e^{\lambda t}.$$

Here, the amplitudes of the perturbation, A and B, are small compared to the amplitude of the spatially uniform solution, and we assume an infinitely large space. The equation for the new eigenvalue, λ, is

$$\det \begin{pmatrix} \alpha - D_c\omega^2 - \lambda & -\beta \\ a_I & -b_I - D_I\omega^2 - \lambda \end{pmatrix} = 0, \tag{9.11}$$

where we define

$$\alpha = \frac{\hat{C} r a_p}{b_p(1+\epsilon\hat{C})^2(1+\hat{I})} > 0, \quad \beta = \frac{\hat{C}^2 r a_p}{b_p(1+\epsilon\hat{C})(1+\hat{I})^2} > 0.$$

Equation (9.11) can be written as

$$\lambda^2 + \lambda(b_I - \alpha + (D_C + D_I)\omega^2) + a_I\beta - (b_I + D_I\omega^2)(\alpha - D_C\omega^2) = 0. \quad (9.12)$$

This is the dispersion relation which connects the growth-rate, λ, with the spatial frequency of the perturbation, ω. The stability conditions are now given by

$$b_I - \alpha + (D_C + D_I)\omega^2 > 0, \quad (9.13)$$

$$a_I\beta - (b_I + D_I\omega^2)(\alpha - D_C\omega^2) > 0. \quad (9.14)$$

Note that the stability conditions for solution $(\hat{C}, \hat{I})$ of the system of ODEs, (9.4-9.5), are obtained automatically from the conditions above by setting $\omega = 0$:

$$b_I - \alpha > 0, \quad (9.15)$$

$$a_I\beta - b_I\alpha > 0. \quad (9.16)$$

Inequality (9.13) is always satisfied because of inequality (9.15). Let us derive conditions under which the spatially uniform solution is unstable. This requires that condition (9.14) is reversed. This can be expressed as follows:

$$F(\omega) \equiv D_I D_C \omega^4 - \omega^2\gamma_1 + \gamma_2 < 0, \quad (9.17)$$

where we denoted for simplicity,

$$\gamma_1 = \alpha D_I - b_I D_C, \quad \gamma_2 = a_I\beta - \alpha b_I > 0.$$

This is a fourth order polynomial, symmetrical with respect to the line $\omega = 0$, with a positive leading term. The points, $\pm|\omega|$, satisfying

$$\omega^2 = \frac{\gamma_1}{2 D_I D_C}, \quad (9.18)$$

correspond to the two minima of the left hand side of inequality (9.17). Let us call these values of ω, $\pm\omega_c$. The condition $F(\omega_c) < 0$ defines that the uniform solution $(\hat{C}, \hat{I})$ is unstable.

Let us plot the function $F(\omega)$ for different values of a_I, see Figure 9.5. For small values of a_I, $F(\omega)$ is strictly positive, and the spatially uniform

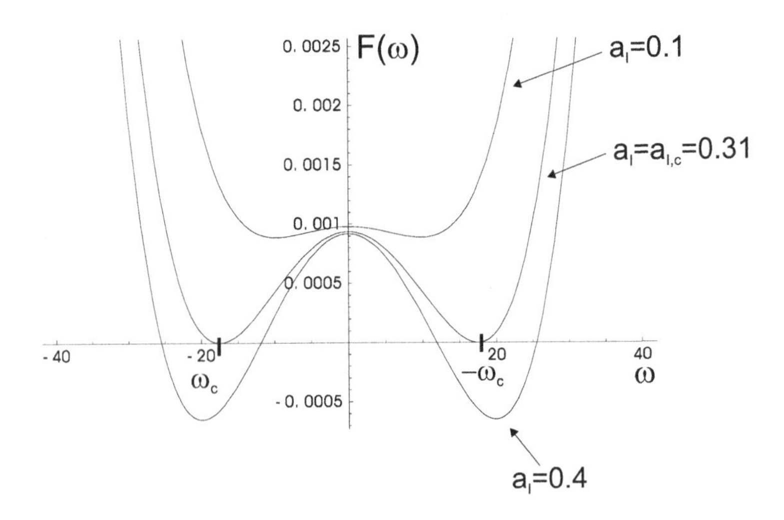

Fig. 9.5 Emergence of Turing instability. As a_I increases and through its critical value, the function $F(\omega)$ (equation (9.17)) crosses zero. Negative regions of $F(\omega)$ correspond to unstable wave-numbers. The wave-number which becomes unstable first is denoted by ω_C. The parameters are as follows: $r = 1; \delta = 0.1; a_P = 5; b_P = 0.1; b_I = 0.01; D_C = 0.00001; D_I = 0.001$.

solution is stable. As a_I increases, the function $F(\omega)$ crosses the line $F = 0$. The critical value of a_I, $a_{I,c}$, for which $F(\omega_c) = 0$, is determined from

$$(\alpha D_I - b_I D_C)^2 = 4 D_I D_C (a_I \beta - \alpha b_I),$$

where α and β both depend on a_I. We solved this equation numerically to find the critical value of $a_{I,c}$, see Figure 9.5.

The applicability of the above analysis depends on the parameters of the system. First of all, we need conditions (9.15-9.16) to be satisfied. They mean that without diffusion, a positive, spatially uniform solution is stable. Next, we need to be in a *weakly nonlinear regime*, where the function $F(\omega)$ has only very narrow regions of ω corresponding to negative values. More precisely, $\Delta\omega \sim L^{-1}$, where L is the spatial dimension of the system. In terms of parameter a_I, we require that it is sufficiently close to $a_{I,c}$. Then, we can calculate the "most unstable" wave–number, that is, ω_c defined by equation (9.18), with $\omega_{c,I}$. This value will determine the spatial period of the solution,

$$\text{Period} = \frac{2\pi}{\omega_c}. \tag{9.19}$$

9.3.2 *Stationary periodic solutions*

Let us start from the value a_I below the critical, $a_I < a_{I,c}$. The system exhibits bistability. If we start in the vicinity of a $(0,0)$ solution, then cancer will not grow and decay to zero. If we start from a point (C, I) in the domain of attraction of the solution $(\hat{C}, \hat{I})$, then the system will develop towards this positive spatially homogeneous stationary solution.

Next, let us suppose we have $a_I > a_{I,c}$, but make sure that it is sufficiently close to $a_{I,c}$ (the exact meaning of "close" is specified in the analysis above). Again, if the initial conditions are close to the zero solution, then the zero state will be the state that the system will attain. However, if we start in the vicinity of the $(\hat{C}, \hat{I})$ state, we will observe interesting behavior. Solution $(\hat{C}, \hat{I})$ is now unstable, and we will see "ripples" developing on top of this solution. This is Turing instability. The spatial period of the ripple was calculated in the previous section. Long-time evolution of this state is of course not in the realm of linear stability analysis, but we can predict that the spatial scale of the resulting solution will be given by (9.19).

Finally, let us assume that a_I is much higher than critical. Now, solution $(\hat{C}, \hat{I})$ is unstable even in the system of ODEs. However, a periodic solution will develop, unless the initial condition is in the domain of attraction of the zero solution. The spatial scale of the periodic solution is determined intrinsically by the parameters of the system, and it grows with a_I. Intuitively this is easy to understand, because higher values of a_I correspond to higher levels of inhibition, so the distance between regions of large C will become larger. Note that the exact period of the periodic solution is adjusted to fit the boundary conditions of the system. For instance, with the Neumann boundary conditions, the boundary points are forced to be troughs of the wave-like pattern. In other words, the period of the solution must be an integer fraction of L.

9.3.3 *Biological implications and numerical simulations*

We start with a scenario where the degree of inhibition is much larger than the degree of promotion ($a_I/b_I >> a_p/b_p$). This corresponds to the early stages when the tumor is generated. We then investigate how tumor growth changes as the degree of inhibition is reduced relative to the level of promotion (i.e. the value of a_I/b_I is reduced). We consider the following parameter regions (Figure 9.6).

(1) If the degree of inhibition is strong and lies above a threshold, growth

of the cancer cells to higher levels does not occur (not shown). Only a small number of cells which do not require promotion for survival would remain.

(2) If the degree of inhibition is weaker, the cancer cells can grow. The spread across space is, however, self-limited (Figure 9.6a). The cancer cells migrate across space. The inhibitors produced by the cancer cells also spread across space, while the promoters do not. Therefore, as the cancer cells migrate, they enter regions of the tissue where the balance of inhibitors to promoters is heavily in favor of inhibitors. Consequently, these cells cannot grow within the space. They remain dormant and may eventually die. In biological terms, this corresponds to a single coherent but self-limited lesion (*uni-focal*). Note that this does not mean that it is in principle impossible to generate more lesions. It means that the space between lesions is bigger than the space provided for cancer growth within the tissue.

(3) As the production of inhibitors is further reduced, we enter another parameter region. Now fewer inhibitors diffuse across space. We observe that multiple lesions or foci are formed (Figure 9.6b). They are separated by tissue space which does not contain any tumor cells. The separate lesions produce some inhibitors, and they diffuse across space. This explains the absence of tumor cells between lesions. Because the production of inhibitors is weakened, however, tumor growth is only inhibited in a certain area around the lesion, and not across the whole space. How many lesions are found within a tissue depends on the parameters in the model, in particular on the relative strength of inhibition and promotion (Figures 9.6b and c). The stronger the degree of inhibition, the larger the space between lesions, and the fewer lesions we expect. The weaker the degree of inhibition, the smaller the space between lesions, and the larger the expected number of lesions. Analytical expressions for the space between lesions are given below. In biological terms, the occurrence of multiple lesions within a tissue which arise from a single tumor is often referred to as *multi-focal* cancers.

(4) If the degree of inhibition is further reduced and lies below a threshold, spread of inhibitors is sufficiently diminished such that the tumor cells can invade the entire space and tissue (Figure 9.6d). In biological terms, this corresponds to the most extensive tumor growth possible within a tissue.

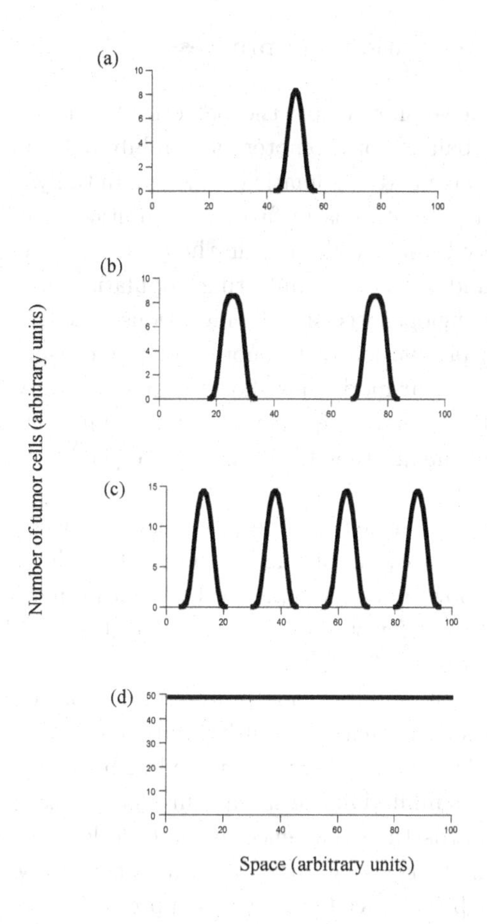

Fig. 9.6 Outcome of the spatial model depending on the relative balance of promoters and inhibitors,captured in the variable a_i. Parameters were chosen as follows: $r = 1; \delta = 0.1; a_P = 5; b_P = 0.1; b_I = 0.01; D_C = 0.00001; D_I = 0.001; L = 2$ For (a) $a_I = 3$, (b), $a_I = 2$, (c), $a_I = 1$, (d) $a_I = 0.1$

In summary, as the relative degree of inhibition is reduced, the patterns of tumor growth change from absence of significant growth, to a single self-limited tumor, to the occurrence of multiple foci, and to the maximal invasion of the tissue by tumor cells. Multi-focal cancers may arise through the dynamical interplay between long range inhibition and local promotion. The following section will examine this in the light of somatic evolution.

9.4 Somatic cancer evolution and progression

The previous sections have shown how the pattern of cancer growth can depend on the relative balance of promoters and inhibitors. Here we consider these results in the context of somatic evolution. Initially, the balance between inhibitors and promoters is in favor of inhibition. Inhibitors are likely to be produced by healthy cells (e.g. in the context of angiogenesis), and they are more abundant than an initiating population of transformed cells. In the context of angiogenesis, specific mutations have been shown to result in the enhanced production of promoters or reduced production of inhibitors in cancer cells. Our model has shown that such mutants have to be produced at a relatively high frequency, so that a sufficient number of promoting cells are present in order to ensure that enough promoters are produced to overcome the effect of inhibition.

Once the promoting cells have succeeded to expand, cancer progression can occur in a variety of ways according to the model. How the cancer progresses depends on how much the balance between promotion and inhibition has been shifted in favor of promotion. We distinguish between three possibilities (Figures 9.7, 9.8 & 9.9).

(i) The balance between inhibition and promotion has been shifted only slightly in favor of promotion, such that self-limited growth of the cancer is observed (Figure 9.7). That is, we observe a single lesion which can grow to a certain size but which is limited in the spread through the tissue. In order to progress further towards the occurrence of multiple lesions or towards more extensive invasion of the tissue, further mutants have to be generated which are characterized by enhanced production of promoters or by reduced production of inhibitors. This introduces a new problem: such a mutant will not have a selective advantage, but is selectively neutral relative to the other cells. This is because the promoters and inhibitors secreted from one cell affect the whole population of cells. If the mutant produces more promoters, not only the mutant, but the entire population of tumor cells benefits. This means that a mutant characterized by enhanced production of promoters will not invade the tumor cell population. Instead, we observe genetic drift which is stochastic and not described by the equations considered here. The model does, however, suggest the following (Figure 9.7): if the population of mutant cells remains below a given threshold relative to the rest of the tumor cells, it will not alter the growth pattern. If the population of mutant cells grows beyond a threshold relative to the rest of the tumor cells, it can change the pattern of cancer growth, even if the mutants do not

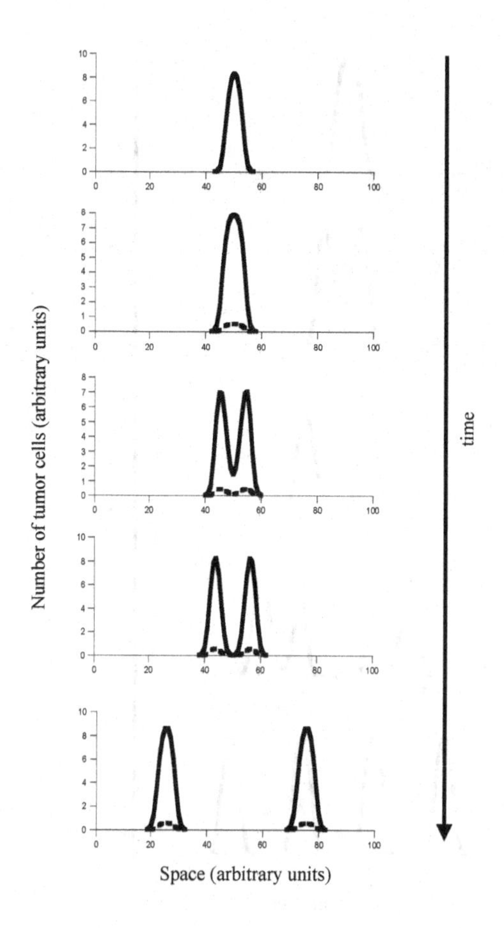

Fig. 9.7 Tumor progression if the initial mutant cell line has only shifted the balance between promoters and inhibitors slightly in favor of promotion. This cell line can only give rise to self limited growth. Further tumor growth requires the generation of further mutants. The new mutant in the simulation is depicted by the dashed line. Parameters were chosen as follows: $r = 1; \delta = 0.1; a_P = 5; b_P = 0.1; a_I = 3; b_I = 0.01; D_c = 0.00001; D_I = 0.001, L = 2$. For mutant: $a_I = 0.5; a_P = 20$.

become fixed in the population (Figure 9.7). The change can either be the generation of multiple lesions, or invasion of the whole tissue, depending on the amount by which the level of promotion has been enhanced by the mutant cell population. The chances that the mutant cell population drifts to levels high enough to cause such a change in tumor growth depend on the population size of the lesion. The larger the number of tumor cells,

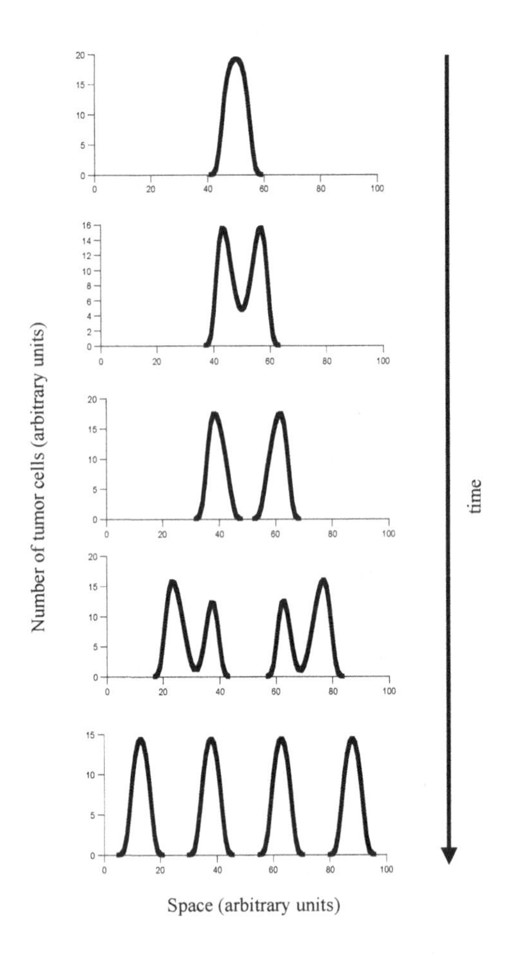

Fig. 9.8 Tumor progression if the initial mutant cell line has shifted the balance between promoters and inhibitors more substantially towards promotion. Now, multiple foci can develop without the need for further mutations. The multiple foci develop, however, by first generating a single lesion which subsequently splits to give rise to two lesions during the natural growth process. Parameters were chosen as follows: $r = 1; \delta = 0.1; a_P = 5; b_P = 0.1; a_I = 1; b_I = 0.01; D_C = 0.00001; D_I = 0.001, L = 2$.

the lower the chance that the relative population size of the mutants can cross this threshold. If this cannot occur, further cancer progression not only requires the generation of a mutation which enhances the level of promotion, but an additional mutation which gives the promoter mutant a selective advantage over the rest of the cell population. That is, in addition to the mutation which shifts the balance in favor of promotion, a mutation is required either in an oncogene or a tumor suppressor gene so that the mutant can grow to sufficiently high numbers or fixation.

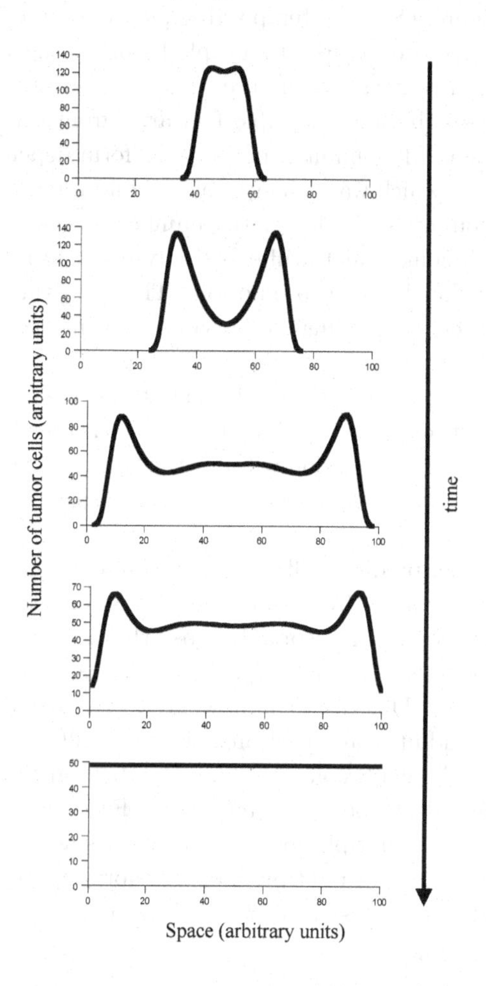

Fig. 9.9 Tumor progression if the initial cell line has largely escaped inhibition, and promotion is the dominant force. Now the tumor grows in space as a single lesion until the whole tissue is invaded. Parameters were chosen as follows; $r = 1; \delta = 0.1; a_P = 5; b_P = 0.1; a_I = 0.1; b_I = 0.01; D_C = 0.00001; D_I = 0.001, L = 2$.

(ii) The first mutation shifts the balance between promoters and inhibitors to a lager extent which is sufficient to result in the generation of multiple lesions (Figure 9.8). The multiple lesions do not, however, occur immediately. First, the tumor grows as a single and self limited lesion (Figure 9.8). Over time, this lesion bifurcates to give rise to two lesions, or further lesions if the degree of promotion is large enough relative to the degree of inhibition (Figure 9.8). The temporal sequence from a single and self-controlled lesion to the occurrence of multiple lesions is the same as in the previous case. But in contrast to the previous case, no further mutations are required. This is because multiple foci arise from the split and migration of a single lesion. The number of foci that form depends on the exact degree of promotion which was achieved by the initial mutation. The higher the degree of promotion, the larger the number of lesions. Growth beyond this number of lesions (which will eventually result in maximal invasion) then requires higher levels of promotion. This is in turn achieved by further mutational events according to the same principles as described in the previous section.

(iii) Finally, assume that the initial mutation shifts the balance so much in favor of promotion that maximal invasion of the tissue is possible (Figure 9.9). Now we observe cancer progression without the generation of multiple foci. Instead, a relatively small single lesion expands in space until all the tissue has been invaded.

In summary, the model predicts different modes of cancer progression in relation to the evolution away from tumor inhibition and towards promotion. A single cancer lesion may spread across the tissue without the occurrence of multiple lesions. Alternatively, the cancer can first grow as a single, self-contained lesion. This can then bifurcate to give rise to multiple foci, either as a result of additional mutations, or as a result of the natural pathway by which multiple foci are generated, depending on the degree of tumor promotion conferred by the initial mutation. Further evolutionary events can then induce the multiple foci to become a single, maximally invasive mass. The occurrence of multiple foci therefore represents an intermediate stage in tumor progression towards malignancy.

9.5 Clinical implications

The occurrence of multiple lesions is observed in a variety of cancers. That is, not one, but several lesions are observed within a given tissue. Multiple lesions can occur by two basic mechanisms [Hafner *et al.* (2002); Hartmann *et al.* (2000); Ruijter *et al.* (1999); Tsuda and Hirohashi (1995); Wilkens *et al.* (2000)]. Either they originate independently by separate carcinogenic events, or they are generated by a single transformation event (monoclonal origin). Sometimes, the term "multi-centric cancers" is used to describe the occurrence of clonally unrelated lesions, while the term "multi-focal" refers to a monoclonal origin [Teixeira *et al.* (2003)]. Clinically, it is important to determine the nature of multiple lesions. The occurrence of multiple lesions can be indicative of a familial cancer, especially if they occur at a relatively young age. Examples are familial adenomatous polyposis (FAP) in the colon, and familial retinoblastoma [Marsh and Zori (2002)]. The genetic predisposition of such individuals renders multiple independent carcinogenic events likely. Alternatively, multiple independent lesions can arise because a large area of tissue has been altered and is prone to the development of cancer, such as Barrett's esophagus [van Dekken *et al.* (1999)], or by other mechanisms which are not yet understood. On the other hand, genetic analysis has indicated that multiple lesions in several cases have a monoclonal origin [Antonescu *et al.* (2000); Holland (2000); Junker *et al.* (2002); Kupryjanczyk *et al.* (1996); Louhelainen *et al.* (2000); Middleton *et al.* (2002); Miyake *et al.* (1998); Noguchi *et al.* (1994); Rosenthal *et al.* (2002); Simon *et al.* (2001); van Dekken *et al.* (1999)]. Examples are mammary carcinoma, gliomas, renal cell carcinoma, hepatocellular carcinoma, and esophageal adenocarcinoma.

The models discussed here show that multiple foci with a monoclonal origin can develop through a dynamical interplay between tumor promoters and inhibitors. The cancer can only grow to high loads as a single mass if it has largely escaped all inhibitory effects. Otherwise, the cancer is likely to grow via the generation of a relatively small and self limited tumor which then bifurcates into multiple foci until it finally invades the entire tissue. The occurrence of multiple foci is therefore an intermediate stage in cancer progression. The higher the number of foci, the further advanced the stage of cancer progression.

A clinically important step in carcinogenesis is the process of metastasis. That is, the spread of tumor cells to the lymph node, entry into the blood supply, and the spread to other tissues. Various studies have investigated the metastatic potential of multi-focal compared to uni-focal cancers [Andea *et al.* (2002); Junker *et al.* (1999); Junker *et al.* (1997)]. In uni-focal cancers, tumor size has been found to be a predictor of metastatic potential. For staging multi-focal breast carcinomas, it has been suggested to use the diameter of the largest tumor only [Andea *et al.* (2002)]. This, however, assumes that the other foci do not significantly contribute to tumor progression. According to our arguments, this would under-stage the cancer. According to the model, the number of foci correlates with the stage of the disease. This has also been concluded in clinical studies, and is supported by data which show reduced patient survival with multi-focal compared to uni-focal cancers [Andea *et al.* (2002)]. Moreover, because our model suggests that multi-focality can occur as a result of reduced tumor cell inhibition, successful metastatic growth might be easier to achieve. Although under debate, some data suggest that inhibitors produced by the primary tumor can prevent metastatic cells from growing [Ramanujan *et al.* (2000)]. If multi-focality correlates with reduced inhibition, then it could also correlate with an increased chance that metastatic cells grow and do not remain dormant.

Further, it is important to note that studies which aim to assess the correlation between multi-focality and metastatic potential should not only concentrate on the number of foci, but also on the size of the foci. As we have shown with the model, cancer progression might start with a small single lesion which can be considered uni-focal. It can then bifurcate to give rise to multiple foci, and finally spread through the entire tissue. When such spread occurs, the multiple foci turn into a big and single mass, and this would again be considered uni-focal. Hence, the cumulative size or volume of the tumor is likely to be the best predictor of malignant progression.

Chapter 10

Mechanisms of tumor neovascularization

In Chapter 9 we explored general models of tumor inhibition and promotion and investigated how the requirement for blood supply influences tumor initiation and progression. The models did not make any specific assumptions about the mechanism by which the tumor induces the formation of new blood supply. Here, we will address this question. Blood vessels are built from so–called *endothelial cells*. As explained in Chapter 9, new endothelial cells are generated and form blood vessels if the balance between promoting factors and inhibiting factors is in favor of promotion. So far, there are two basic hypotheses regarding the mechanism by which tumors induce the generation of new blood vessels.

Traditionally, it was thought that promoters induce local endothelial cells to divide. That is, endothelial cells which make up pre-existing blood vessels divide and give rise to new endothelial cells. These new endothelial cells build more blood vessels. This process has been termed angiogenesis.

Recently, another mechanism has been suggested. According to this hypothesis, the promoting factors induce a circulating population of endothelial progenitor cells (EPCs) to migrate to the site of the tumor and to build new blood vessels locally. The progenitor cells are primitive stem cells which will differentiate into endothelial cells. This is in contrast to the previous mechanism where new endothelial cells were derived from the division of already existing and fully differentiated endothelial cells. There is some experimental evidence which supports a role for progenitor cells in the formation of blood vessels in cancer [Asahara *et al.* (1999); Bolontrade *et al.* (2002); Drake (2003); Rabbany *et al.* (2003); Ribatti *et al.* (2003)]. In order to distinguish this mechanism from the previous one, we will refer to it as vasculogenesis (derived from the process of post-natal vasculogenesis). Note, however, that this might not be a universally accepted term.

In fact, the use of progenitor cells for the generation of blood vessels is sometimes referred to as angiogenesis in the literature. We use the term vasculogenesis only for the purpose of distinction.

Given that there are two possible ways to build blood vessels, which one is more important in the context of cancer? How do the two mechanisms influence the pattern of tumor growth? This Chapter discusses mathematical models which have investigated these questions. They have given rise to suggested experiments which can determine the relative importance of angiogenesis versus vasculogenesis.

10.1 Emergence of the concept of postnatal vasculogenesis

The term "vasculogenesis" was first clearly defined and opposed to the term "angiogenesis" by W. Risau [Risau *et al.* (1988)]. In his classic 1997 *Nature* review he used the assumption that vasculogenesis occurs only during embryonic life [Risau (1997)]. Indeed at that time, there was no direct evidence for postnatal vasculogenesis. The assumption that vasculogenesis occurs only during the embryonic period still persists in some academic circles, as well as in text books on histology. It is argued that once the vascular endothelial system is formed, angiogenesis becomes the predominant mechanism of vascular regeneration during wound healing, as well as during cyclic (physiological) and pathological postnatal vascular morphogenesis.

The concept of postnatal vasculogenesis started emerging in the second half of the 1990s. In 1995, in his review paper in *Nature Medicine* Judah Folkman wrote: "Postnatal vasculogenesis has never been observed, but it would not be entirely surprising if it were discovered in tumors." [Folkman (1995a)]. The situation changed dramatically after the appearance of the paper by Jeffrey Isner and colleagues in *Science* about identification, isolation and angiogenic potential of circulated endothelial progenitor cells [Asahara *et al.* (1997)]. It was the first publication presenting clear evidence of postnatal vasculogenesis. Since then, a number of publications on postnatal and tumor vasculogenesis have appeared. The evidence keeps growing, and there are already several excellent reviews in this field [Asahara *et al.* (1999); Bolontrade *et al.* (2002); Drake (2003); Rabbany *et al.* (2003); Ribatti *et al.* (2003)].

10.2 Relative importance of angiogenesis versus vasculogenesis

There are only three logically possible situations reflecting the relationship between tumor vasculogenesis and angiogenesis.

(1) Tumor angiogenesis exists but tumor vasculogenesis does not exist. This was the dominant view before the ground breaking paper by Asahara [Asahara *et al.* (1997)].
(2) Tumor vasculogenesis is the only mechanism of tumor vascular morphogenesis; angiogenesis does not play any role. Existing experimental data contradict this hypothesis.
(3) Tumor vasculogenesis and tumor angiogenesis coexist. This is what we assumed in the model. There are several ways in which the two processes can co-occur.
   (a) In different types of tumors, the relative contribution of vasculogenesis and angiogenesis to tumor vascular morphogenesis is different;
   (b) This relation depends on patients' age;
   (c) This relation changes during the dynamics of tumor growth and depends on the stage of tumor growth;
   (d) There are tissue-specific and organ-specific differences in the relationship between angio- and vasculogenesis in tumors;
   (e) The relative roles of tumor angiogenesis and vasculogenesis can vary inside the tumor.

Finally, one can imagine very complex combinations of all of the above factors. In fact, it is probably safe to predict that in reality, we are dealing with some sort of a combination of many components. It is obvious that this question is the subject of future intensive research. Here we pursue the following strategy. We fist assume that the process of angiogenesis dominates, and use a mathematical model to determine the patterns of typical tumor growth dynamics. Then, we make the opposite assumption and describe the growth of tumor dominated by vasculogenesis. We show that vasculogenesis-driven and angiogenesis-driven tumors grow in different ways. Once we know this, we can use data to identify the "signature" of these processes by measuring relevant variables, such as tumor growth, the level of circulating stem cells, the state of the bone marrow, etc. This knowledge will help to identify the relative contributions of the two processes in cancer progression.

10.3 Mathematical models of tumor angiogenesis and vasculogenesis

The model describes interactions between three compartments, the bone marrow, the blood and the tumor vasculature, see Fig. 10.1. Let us denote the number of endothelial progenitor cells (EPC) in the bone marrow (BM) at time t as $x(t)$, the number of EPC circulating in the blood system as $y(t)$, and the number of cells involved in the tumor vasculature as $z(t)$.

We will consider two mechanisms by which the tumor's vasculature is built [Komarova and Mironov (2004)]:

(i) angiogenesis,
(ii) vasculogenesis.

Tumor vascular cells that originate (or are descendants of cells that originate) by means of mechanism (i) are denoted as z_a. The ones that come about by means of mechanism (ii) are denotes as z_v. The subscripts in z_a and z_v refer to *angiogenesis* and *vasculogenesis* respectively. The total number of cells involved in tumor vasculature is simply given by $z = z_a + z_v$. In Figure 10.1 we schematically denote angiogenesis-derived cells as white, and vasculogenesis-derived cells as gray. Tumor vasculature will consist of a mixture of the two types of cells.

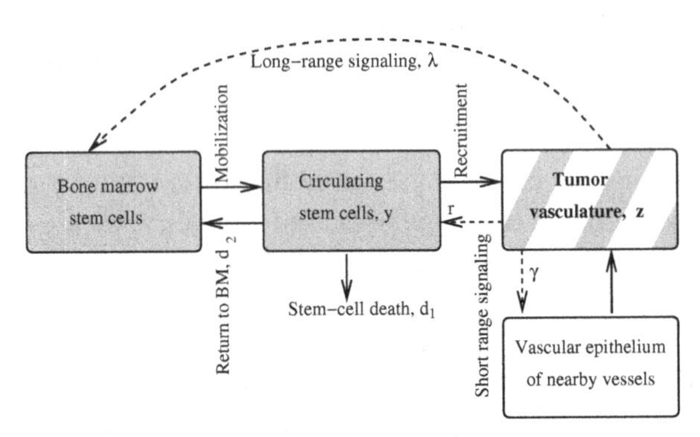

Fig. 10.1 Angiogenesis- and vasculogenesis-related formation of tumor vasculature.

Let us first consider mechanism (i). As the tumor grows, it excretes tumor angiogenesis factors, or TAF, that help activate the endothelial cells of nearby blood vessels. Some tumors, such as many gliomas, secrete the vascular endothelial growth factor (VEGF) that is normally produced by kidneys and brain cells [Plate *et al.* (1992); Shweiki *et al.* (1992)]. The inhibition of VEGF induced angiogenesis suppresses tumor growth in mice [Buchler *et al.* (2003); Kim *et al.* (1993)]. We assume that the degree to which endothelial cells are induced to multiply is proportional to the tumor mass, $M(t)$, which in turn depends on the amount of vasculature, $z(t)$.

Mechanism (ii) involves the following components. We assume that the stem cells suitable for vasculogenesis are supplied to the blood by bone marrow (BM). In the absence of a tumor (or other sites that use circulating stem cells), they enter the blood flow with a constant rate, λ_0. They circulate in the blood system and die with a constant rate, d_1, or can return to the BM with the rate, d_2. If there is no tumor (or any other need for recruitment) then there is a constant, steady concentration of stem cells in the blood [Rafii *et al.* (2003)]. In the presence of a tumor, BM stem cells are mobilized into the blood, by means of *long-range* signaling, mediated by granulocyte-macrophage colony stimulating factor (GM-CSF). This means that the number of stem cells delivered into the blood flow by the BM is increasing with the rate proportional to the tumor size. Experimental evidence of this mechanism is available in the literature, see e.g. [Takahashi *et al.* (1999)], [Shirakawa *et al.* (2002)]. On the other hand, the tumor recruits stem cells from the blood by means of short range signaling (which involves cell adhesion molecules, [Joseph-Silverstein and Silverstein (1998)]). The vascular endothelium of the nearby vessels becomes activated and allows the stem cells to extravascate and start a cycle of differentiation/division. The rate at which activation proceeds is also proportional to the tumor load (which in turn is proportional to z).

Finally, we assume that both the newly stimulated epithelial cells (mechanism (i)) and the recruited stem cells (mechanism (ii)), enter a stage of clonal expansion and continue to form blood vessels. The law of growth is chosen to be consistent with the following simple mechanism: the new blood vessels mostly form on the surface of the growing tumor. This means that the rate of growth is proportional to the tumor surface, $S(t)$. For the time-scales of interest, this is not an unreasonable assumption. Saturation of growth (due to lack of space or other constraints) happens much later and is not considered here. What we would like to calculate is the rate at which a tumor can grow, the relative contributions of the two mechanisms, and how this changes over time.

The assumptions listed above lead to the following system of ordinary differential equations,

$$\dot{x} = b_0 - \lambda_0 x - \lambda M x - d_0 x + d_2 y, \tag{10.1}$$

$$\dot{y} = \lambda_0 x + \lambda M x - (d_1 + d_2) y - rSy, \tag{10.2}$$

$$\dot{z}_a = \gamma S, \tag{10.3}$$

$$\dot{z}_v = rSy. \tag{10.4}$$

All the variables and parameters are summarized in Table 5.1. We expect the tumor mass to grow as a function of vasculature; in general we assume that

$$\text{Tumor mass} \propto (\text{Tumor vasculature})^a, \tag{10.5}$$

where a is some positive number. An example of a possible power law is given by $a \approx 1$, if vasculature is distributed throughout the body of the tumor. If we assume a fractal structure of vasculature, this exponent may be different. The main assumption here is that the tumor mass adjusts instantaneously to the growing size of the vasculature. This is a quasistationary approach. In a more general scenario, one could introduce some rate at which the tumor mass adjusts to changes in vasculature.

With $a = 1$ in equation (10.5), equations (10.3) and (10.4) can be replaced by

$$\dot{M}_a = \gamma S, \tag{10.6}$$

$$\dot{M}_v = rSy. \tag{10.7}$$

We have $M = M_a + M_v$. Depending on the tumor geometry, the relationship between the tumor mass and its surface changes. The basic formula is $S = DM^{\frac{D-1}{D}}$, where D is the dimension of the tumor. For instance, in flat tumors, such as bladder carcinoma in situ, certain cancers of the eye or flat adenomas/adenocarcinomas of the colorectal mucosa [Hurlstone *et al.* (2004); Rubio *et al.* (1995)] we have $D = 2$. The two-dimensional analysis is also applicable in certain *in vitro* experiments. Most of the time, however, solid tumors are three-dimensional, and we will set $D = 3$.

10.4 Mathematical analysis

Let us rewrite the system in a closed form,

$$\dot{x} = b_0 - \lambda_0 x - \lambda M x - d_0 x + d_2 y, \tag{10.8}$$

$$\dot{y} = \lambda_0 x + \lambda M x - (d_1 + d_2) y - rSy, \tag{10.9}$$

$$\dot{M} = \gamma S + rSy, \tag{10.10}$$

with $S = DM^{\frac{D-1}{D}}$.

Tumor-free equilibrium. In the absence of a tumor, we have $M_a = M_v = M = S = 0$, and the equilibrium level of circulating stem cells is determined by setting the right hand side of equations (10.8), (10.9) to zero, which gives

$$x(0) = x_0 = \frac{b_0(d_1 + d_2)}{d_0(d_1 + d_2) + d_1 \lambda_0}, \quad y(0) = y_0 = \frac{b_0 \lambda_0}{d_0(d_1 + d_2) + d_1 \lambda_0}.$$

If the rate of return to BM is negligible, $d_2 \ll d_1$, then these equilibrium expressions become more transparent,

$$x(0) = x_0 = \frac{b_0}{d_0 + \lambda_0}, \quad y(0) = y_0 = \frac{\lambda_0 x_0}{d_1}.$$

The expression for x_0, the equilibrium number of BM EPCs in the absence of a tumor, is a balance between a constant production, tumor-independent recruitment and death. We conclude that in the absence of tumor,

$$\text{BM EPCs} = \frac{\text{Constant production}}{\text{Constant Recruitment} + \text{Death}}.$$

This is an unstable equilibrium. Adding a small amount of M will get it out of balance and lead to a growth of tumor. It is this process that we will model next. We start from the initial condition

$$M = \epsilon, \quad x(0) = x_0, \quad y(0) = y_0.$$

The advantage of the above system is that we can consider the two regimes, angiogenesis and vasculogenesis, separately. Namely, by taking $\gamma = 0$, we can assume that the only way the tumor vasculature is built is by vasculogenesis. Alternatively, $r = \lambda = 0$ means that we assume that the only mechanism is angiogenesis.

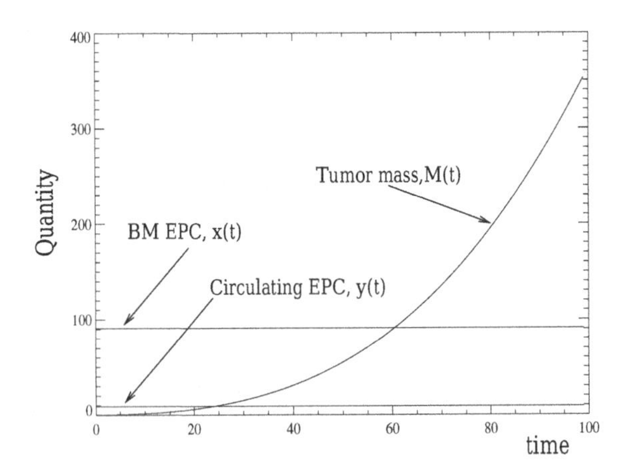

Fig. 10.2 Typical angiogenesis-driven tumor dynamics: the quantities $x(t)$, $y(t)$ and $M(t)$ are plotted as a function of time. The parameter values are as follows: $D = 3$, $b_0 = 10$, $\lambda_0 = 0.01$,$\lambda = 0$, $d_0 = 0.1$, $d_1 = 0.1$, $d_2 = 0$, $r = 0$,$\gamma = 2$.

Angiogenesis-driven dynamics. We start with $r = \lambda = 0$. Let us suppose that the linear size of tumor is a. Then $M \propto a^D$ and $S \propto Da^{D-1}$. From the equation (10.10) with $r = 0$, we have

$$\dot{a} \propto \gamma,$$

and we have the law for tumor growth,

$$M(t) \propto t^D,$$

which is valid for all times. Obviously, in this regime, the equation for the tumor becomes independent of the equation for circulating EPCs. The numbers of the BM EPCs and circulating EPCs remain at their equilibrium level,

$$x(t) = x_0, \quad y(t) = y_0.$$

The very simple behavior of this system is presented in Figure 10.2.

Vasculogenesis-driven dynamics. Next, let us consider the opposite regime by taking $\gamma = 0$. Now all the equations are coupled. As time goes by, the tumor size will increase, and this will lead to an increase in the number of circulating EPCs. Consequently, the number of BM EPCs will drop. Qualitatively, the dynamics can be described as follows. In the beginning, $y(t)$ increases. Then it reaches a maximum, after which it will decrease to zero. $x(t)$ drops to low numbers as $y(t)$ reaches its maximum, and then it continues to decrease to zero. The dynamics of $M(t)$ also has two stages. It grows faster than linear in the beginning, and then saturates at a linear growth with slope b_0, as $y \to \infty$.

To understand the long-term behavior, let us suppose that $d_2 = 0$, since it does not change the behavior qualitatively. We note that initially, the balance in the equation for x is defined by the b_0 and $-(\lambda_0 + d_0)x$ terms. As time goes by and M increases, we have the balance $b_0 \approx \lambda M x$, which means that $x \to 0$. Also, this expression defines the dynamics of y; we have $b_0 \approx rSy$, where $S \to \infty$, and $y \to 0$.

A typical outcome of a numerical simulation of system (10.8-10.10) is given in Figure 10.3. A note of caution: the numerical values of the functions should not be taken literally. The point of Figures 10.2 and 10.3 is to depict the qualitative behavior of the angiogenesis- and vasculogenesis-driven systems, which turns out to be very different.

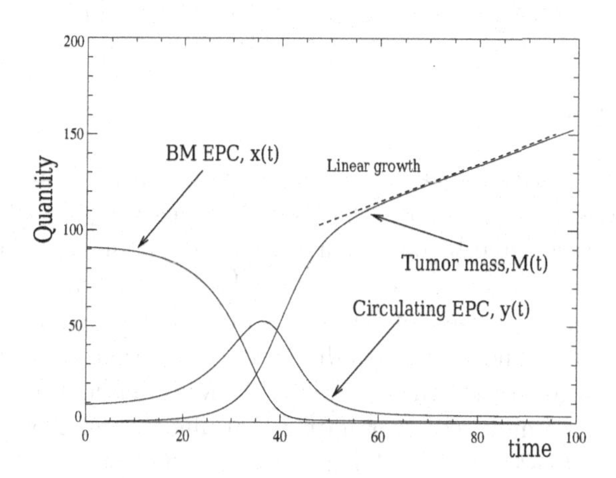

Fig. 10.3 Typical vasculogenesis-driven tumor dynamics: the quantities $x(t)$, $y(t)$ and $M(t)$ are plotted as a function of time. The parameter values are as in Figure 10.2, except $\lambda = 0.1$, $r = 0.1$ and $\gamma = 0$.

Main qualitative differences. From this analysis we conclude that the behavior of the system driven by angiogenesis and vasculogenesis is different. The two main points of difference are as follows:

- For angiogenesis driven systems, the amount of BM EPCs and circulating EPCs stays constant in time. For vasculogenesis driven systems, the amount of BM EPCs steadily decreases, and the amount of circulating EPCs experiences a sharp peak in the beginning and then also decreases.
- The tumor mass in angiogenesis driven systems grows as a cube of time, that is, the diameter of the tumor grows linearly with time. For vasculogenesis driven systems, tumor growth has two stages: at first, the tumor mass grows faster than linear, and later, once the BM is depleted of EPCs, the tumor mass grows linearly with time, which means that the diameter of the tumor grows as a cubic root of time.

Another mathematical model of tumor vasculogenesis has been recently proposed [Stoll *et al.* (2003)]. In this paper, the emphasis is on the geometry of the tumor and its growth dynamics. However, this model does not take account of the independent dynamics of BM and circulating EPCs. Our model concentrates on the description of the fine balance between the three compartments: the BM, the circulatory system and the tumor.

10.5 Applications

We have presented a mathematical model of tumor growth driven by angiogenesis and vasculogenesis and found that the dynamics are quite different in the two cases, as predicted by our equations. Indeed, if angiogenesis was the entire story, then we would expect a cubic (third power) growth of the tumor mass, and constant levels of BM EPCs and circulating EPCs. On the other hand, if the tumor growth is driven by vasculogenesis, then the dynamics will go through two stages. First, the level of circulating EPCs will increase and the tumor will grow faster than linear, and then, when the BM is depleted of EPCs, the level of circulating EPCs will also go down and tumor growth will slow down to linear, which means that the tumor diameter will only grow as a cubic root of time.

10.5.1 *Dynamics of BM-derived EPCs*

Even though the exact role the BM-derived EPCs play in the formation of *de novo* blood vessels in the process of tumorigenesis is heavily debated, there is growing evidence of their importance for the formation of tumor vasculature [Bolontrade *et al.* (2002); Davidoff *et al.* (2001)]. A paper by [Lyden *et al.* (2001)] suggests that "recruitment of VEGF-responsive BM-derived precursors is necessary and sufficient for tumor angiogenesis". Schuch et al. [Schuch *et al.* (2003)] propose that EPCs can be a novel target for endostatin and suggest that their relative numbers can serve as a surrogate marker for the biological activity of antiangiogenic treatment.

In this chapter, we developed a model with a predictive power regarding the dynamics of BM-derived EPCs. This is a first attempt to quantify the level of EPCs in relation to carcinogenesis.

10.5.2 *Re-evaluation of apparently contradictory experimental data*

The two-stage dynamics characteristic of vasculogenesis-driven tumor growth is consistent with some experimental data published recently. In particular, our model can help resolve some contradicting reports on the levels of circulating EPCs in cancer patients. In the paper [Beerepoot *et al.* (2004)] it is found that the level of circulating endothelial cells in peripheral blood of cancer patients is increased compared with healthy subjects. More specifically, cancer patients with progressive disease had on average 3.6-fold more circulating EPCs than healthy subjects. Patients with stable disease had circulating EPC numbers equal to that in healthy subjects. On the other hand, [Kim *et al.* (2003)] reports that the number of circulating EPCs was not found to be increased in cancer patients, although the plasma levels of VEGF were elevated. It was further concluded by the authors that VEGF, at concentrations typical of those observed in the blood of cancer patients, does not mobilize EPCs into the peripheral blood.

With our model, it is possible to resolve this apparent discrepancy in measurements of the level of circulating EPCs. If we look at Figure 10.3, we can see that the level of EPCs first increases and then drops even below the level corresponding to the equilibrium in healthy subjects. Therefore, the *timing* of measurements becomes crucially important. The level of circulating EPCs will depend on the stage of cancer development. It experiences a peak and drops considerably afterward. It would be very interesting to perform systematic measurements to find out the exact timing of this process.

Application for diagnostics. Dynamic analysis of the number of circulated EPCs in blood opens up a clinically important avenue of research. Circulated stem cells can be used as a surrogate marker of tumor vasculogenesis. Development of assays which would allow us to monitor the recruitment of labeled EPCs could eventually be transformed into a clinical diagnostic test for estimating the intensity of tumor vasculogenesis.

10.5.3 *Tumor growth kinetics*

The kinetics of tumor growth is a complicated question, and no universal answer has been given as to how exactly tumors develop in time. There may be a good reason for this: different tumors may grow according to different scenarios, and these scenarios may be very complicated. The *Gompertzian* law of tumor growth has been extensively discussed over the last four decades, see e.g. [Laird (1969); Lazareff *et al.* (1999); Norton (1988)]. This "sigmoidal" empirical law models an exponential growth of a tumor at the initial stages followed by saturation at a constant level. There are many papers which suggest that this law does not hold [Retsky *et al.* (1990)], and propose different models [Ferreira *et al.* (2002); Gatenby and Gawlinski (1996); Guiot *et al.* (2003); Kansal *et al.* (2000); Sherratt and Chaplain (2001)]. Many of the mathematical models seem to be concerned with the *avascular* stage in tumor growth. While this may be of theoretical interest, it is believed that most of the observed tumors are dependent on blood supply. Therefore, the formation of new blood vessels should be a part of a realistic model.

In general, there is a curious situation regarding the state of affairs with the studies of tumor growth kinetics. It seems that many modelers come up with different theoretical constructs predicting various modes of growth, while there is very little directed research in this area on the part of experimental cancer biologists. One possible explanation is of course the impossibility of studying tumor growth kinetics in humans, without treatment. Another factor is the difficulty of precise measurements: with only a limited number of data points and a large error of measurement, it is impossible to make out subtleties of the growth dynamics. Finally, the very complexity of multistage tumorigenesis poses a problem when trying to identify any "universal" behavior, thus leaving theorists with their theories unchecked, and making experimental biologists concentrate on issues of "survival", "treatment success" etc., which present more possibilities for direct applications in treatment.

It is probably safe to say that different factors affect the rate of tumor growth during different stages of tumorigenesis. In the very beginning (the avascular stage), a mutation (or a set of mutations) throws the growth and death regulation out of balance, which may lead to an exponential accumulation of such (pre-)malignant cells. Then, for some reason, the growth slows down. This can be related to space/density control or the lack of specific growth factors. The growth of the lesion plateaus, until the next mutation breaks out of homeostatic regulation, leading to another increase in cell number. At some point, the lesion reaches the size where it is impossible to keep up the functioning of cells unless additional blood supply is provided. At this stage, the rate-limiting factor becomes the making of the new blood vessels.

It is this stage of the growth that we concentrated on in our model. We assumed that new blood vessels are formed near the surface of the existing tumor, thus making the cells near the surface divide more often than the core. A similar assumption was made in the interesting paper by [Bru *et al.* (2003)]. There, a linear growth of the diameter with time was observed in colonies of tumor cell lines *in vitro*. The authors went on to develop a model which takes account of the fractal structure of a tumor. Most of the growth activity (i.e. mitosis) was assumed to be concentrated on the boundary of the colony/tumor, which leads to a linear growth law for the colony diameter.

Our model is similar to this in the assumption that tumor growth happens mostly on the surface. However, it makes more explicit statements on the kinetics of *de-novo* vascularization. If we assume that new blood vessels are formed locally, that is, if the tumor dynamics are dominated by the process of angiogenesis, then we find that tumor mass grows as a third power of time (this means that the tumor diameter grows linearly, like in the model by [Bru *et al.* (2003)]). On the other hand, if circulating EPCs are recruited from the blood stream, that is, vasculogenesis is the dominant process, the growth of the tumor mass (after some transition period) will be linear in time.

The point of our model is to address a specific question, namely, whether new blood vessels are formed "locally" or "globally". It is incomplete without an experimental validation. An experimental test can be performed to find out which of the processes contributes more to tumor growth. If a linear growth is found, then we can conclude that vasculogenesis is more important. If the growth is cubic in time, then angiogenesis wins. Several studies have reported the kinetics of tumor growth which can be used to test our predictions. For instance, in the paper by Schuch et al. [Schuch *et al.* (2002)], the law of tumor volume growth for pancreatic cancers resembles linear, which suggests the prevalence of vasculogenesis. The paper [Hahnfeldt *et al.* (1999b)] contains data on Lewis lung carcinoma implanted in mice, where a linear growth of three-dimensional tumor size was observed. This is again consistent with our vasculogenesis-driven dynamics. On the other hand, [Mandonnet *et al.* (2003)] report the linear growth of the tumor *diameter* for gliomas, which is consistent with the dominance of angiogenesis. However, statistically it may be difficult to distinguish between a linear and a cubic growth curve unless we have many experimental points. A conscious experiment with this specific question in mind would be very desirable to address this issue.

Chapter 11

Cancer and immune responses

As pointed out in the previous chapter, the body is characterized by defense systems which can limit the growth and pathogenicity of selfish tumor cells once they have arisen by a series of mutations. The previous chapter explored how the limitation of blood supply can prevent cancers from growing beyond a very small size and from progressing. This is a mechanism which is supported very well by experimental and clinical data, and which is also studied from a therapeutic point of view. Another mechanism which can potentially counter the growth of cancer cells is the immune system. As will become apparent in this chapter, however, the role of the immune system in cancer is highly debated and uncertain.

The immune system defends human beings from intruders such as pathogens which would otherwise kill them. It does so by specifically recognizing proteins derived from the pathogens (for example, viruses, bacteria, or parasites). Through complicated mechanisms which will be discussed briefly later on, the immune system knows that these proteins are foreign and that they are not derived from the organism that it is supposed to protect. What about cancer? As discussed throughout this book, carcinogenesis involves the accumulation of multiple mutations and in general often exhibits genetic instability. This means that many mutated proteins are produced which are different from the organism's own proteins and should thus appear foreign. In principle, these should be visible to the immune system which could potentially remove tumor cells and prevent the development of cancer.

A role of the immune system in the fight against cancer was first suggested in 1909 by Paul Ehrlich [Ehrlich (1909)]. It was not, however, until the 1950s, when the idea was pursued more vigorously and the immune surveillance hypothesis was formulated by Burnet [Burnet (1957)]. It stated that while cancers continuously arise, they are eliminated by specific immune responses. The successful establishment of cancer was thought to come about by the occasional escape of cancer cells from the immune responses. In support of this hypothesis, experimental data indicated that cancer cells show many characteristics which prevent the immune system from recognizing the mutated proteins and from killing the cells successfully.

Following a lot of enthusiastic research in this context, clinical and experimental data cast doubt on the immune surveillance hypothesis [Dunn *et al.* (2002)]. Patients characterized by impaired immune systems showed no significant increase in the incidence of cancers which are not induced by viruses. Similarly, nude mice – which lack adaptive immune responses – have a similar incidence of cancers compared to normal mice. Moreover, experimental studies tracked immune cells specific for cancer proteins and found that they did not react successfully in the first place. This in fact contradicted the hypothesis that the immune system can play any surveillance role in the context of cancers. Since then, many papers have investigated the relevance of immune responses in cancer [Dunn *et al.* (2002)]. While our understanding is still rudimentary, it seems that the truth lies somewhere in between these two extreme views.

This chapter will review mathematical work which tried to account for some of the experimental data on immunity and cancer. We will start with a brief overview of immunity which gives the necessary background before discussing the model and equations.

11.1 Some facts about immune responses

This section will briefly review some basic immunological principles which form the basis for the rest of the chapter. More extensive descriptions of the immune system can be found in any standard immunology text book, for example [Janeway *et al.* (1999)]. We can distinguish between two basic types of immune responses. *Innate immune responses* provide a first line of defense. They do not recognize foreign proteins specifically. They provide environments which generally inhibit the spread of intruders. While they may be important to limit the initial growth of a pathogen, they are usually not sufficient to resolve diseases. They will not be considered further here. On the other hand, *adaptive immune responses* can specifically recognize foreign proteins and can resolve diseases. We concentrate on this type of response here. Adaptive immunity can be subdivided broadly into two types of responses: antibodies and killer cells. Antibodies recognize proteins outside the cells, such as free virus particles, extracellular bacteria or parasites. Killer cells recognize foreign proteins which are displayed on cells. For example, viruses replicate inside cells. During this process, the cell captures some viral proteins and displays them on the cell surface. When the killer cells recognize the foreign proteins on the cell surface, they release substances which kill. Mutated cancer proteins are displayed on the surface of cancer cells. Therefore, killer cells are the most important branch of the immune system in the fight against cancer. The rest of this chapter will discuss only the role of killer cell responses. The scientific term is *cytotoxic T lymphocyte*, abbreviated as *CTL*. They can also be referred to as CD8+ T cells because they are characterized by the expression of the CD8 molecule on the cell surface.

The CTL are able to recognize the cell which displays a foreign protein in the following way. When proteins inside the cell are captured for display on the cell surface, they are presented in conjunction with so–called *major histocompatibility complex* (MHC) molecules. The MHC genes are highly variable, and different MHC genotypes present different proteins. This accounts for the variability between different people in immune responses against the same pathogen. The CTL carries the so–called T cell receptor or TCR. The TCR recognizes the protein-MHC complexes. This triggers the release of specific molecules such as perforin or FAS, which induce apoptosis in the cell that displays the foreign proteins. In immunology, the foreign protein which is recognized by the immune cells is also referred to as *antigen*.

It is important to note that all proteins of a cell are processed and displayed in this way, not just the ones which are supposed to be recognized by the immune system. With this in mind, how can the immune cells distinguish between self and foreign? After all, the vast majority of proteins produced by a cell are normal self proteins. Immune cells usually do not react against self proteins. This is called *tolerance*. They do, however, react against most pathogens. This is called *reactivity*. In the context of cancers it appears that the immune response can remain tolerant to mutated cancer proteins during the natural history of progression, but that reactivity can be induced by certain vaccination approaches. What determines whether we observe tolerance or reactivity? This question is still debated. Different hypotheses have been put forward. According to one hypothesis, immune cells can distinguish between "self" and "foreign" [Janeway (2002)]. Negative selection early during the development of immune cells can result in the deletion of cells which react to self proteins. This hypothesis has difficulties to explain immunological tolerance in the context of tumors because the mutated tumor cell proteins should appear foreign to the immune system. Another idea is the danger signal hypothesis [Fuchs and Matzinger (1996)]. This states that immune responses only react if they sense so–called "danger signals". They can be released as a result of tissue injury, necrosis, or virus infections. Since tumor cells die predominantly by apoptosis, such danger signals are not released. This could explain why the immune system remains tolerant in the context of cancers, while it reacts to infectious agents. The problem with the danger signal hypothesis is that such signals have yet to be identified in terms of chemical compounds.

Recently, the phenomenon of cross-priming and cross-presentation has received attention in the context of CTL regulation [Albert *et al.* (2001); Albert *et al.* (1998); Belz *et al.* (2002); Blankenstein and Schuler (2002); den Haan and Bevan (2001); den Haan *et al.* (2000); Heath and Carbone (2001a); Heath and Carbone (2001b)]. This relates to the initiation of immune responses. Before the antigen is seen, very few specific immune cells exist. When antigen is recognized the specific CTL start to divide and undergo clonal expansion. If clonal expansion is successful, we observe reactivity. If it is not successful, we observe tolerance. This initiation of the response does not occur at the site where the aberrant target cells are located, but in the lymph nodes. Cross priming means that the initiation of the CTL response is not mediated by antigen which is displayed on the target cells themselves. Instead, it occurs when antigen is recognized on so–called *antigen presenting cells* (APCs). The job of these cells is to take up antigen, transport it to the lymph nodes, and present it to the CTL. This activates the CTL and induces clonal CTL expansion. Once expanded, the CTL can migrate to the site where the aberrant cells are located. Once there, they perform their effector function; that is, they kill the cells.

In summary, the indirect recognition of antigen by CTL on antigen presenting cells is called cross-priming or cross-presentation. On the other hand, the direct recognition of antigen on the target cells which are supposed to be killed is called direct presentation. These concepts are explained in more detail schematically in Figure 11.1.

Because cross-presentation is thought to play a role in deciding whether we observe reactivity or tolerance, this process is the subject of mathematical models in this chapter. We start by introducing a basic mathematical model of CTL response regulation and then discusses basic mathematical results. These mathematical results are subsequently applied to aspects of cancer evolution, progression and treatment.

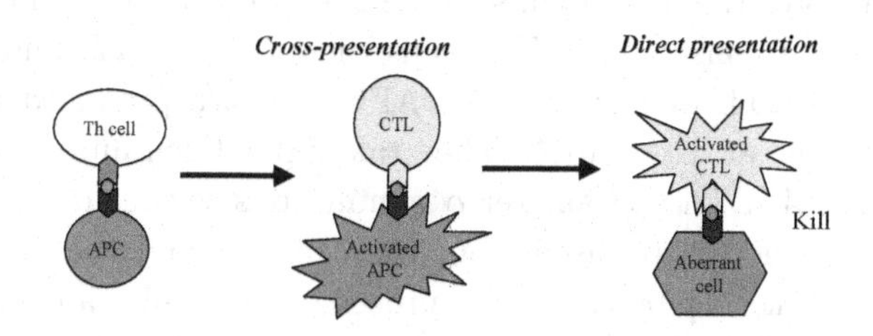

Fig. 11.1 Schematic representation of the concept of cross-priming, which is central to this chapter. So called "antigen presenting cells" can take up antigen (proteins derived from pathogens or cells) and display them on their surface. Before the APCs can function, they need to be activated. This is achieved by so called helper T cells (Th) which can recognize the antigen on the APC. The activated APC subsequently can interact with CTL. CTL can also specifically recognize the antigen on the APC. This interaction activates the CTL which can then turn into effector cells and kill the troubled target cells which display the antigen. These target cells are different from APCs and can be for example virus-infected cells or tumor cells. This process is called cross-priming because the CTL do not get activated directly by the troubled cells which need to be killed, but indirectly by the APCs which can take up and display the antigen.

11.2 The model

We describe a model containing four variables: cells directly displaying antigen such as infected cells or tumor cells, T (we will refer to these cells as "target cells"); non-activated APCs which do not present the antigen, A; loaded and activated APCs which have taken up antigen and display it, A^*; CTL, C. The model is given by the following system of differential equations which describe the development of these populations over time,

$$\begin{aligned}
\dot{T} &= rT\left(1 - \frac{T}{k}\right) - dT - \gamma TC, \\
\dot{A} &= \lambda - \delta_1 A - \alpha AT, \\
\dot{A}^* &= \alpha AT - \delta_2 A^*, \\
\dot{C} &= \frac{\eta A^* C}{\varepsilon C + 1} - qTC - \mu C. \qquad (11.1)
\end{aligned}$$

The infected or tumor cells grow at a density dependent rate $rT(1\text{-}T/k)$. In case of virus infections, this represents viral replication, where virus load is limited by the availability of susceptible cells, captured in the parameter k. In case of tumors, this corresponds to division of the tumor cells, and the parameter k denotes the maximum size the tumor can achieve, limited for example by spatial constraints. The cells die at a rate dT, and are in addition killed by CTL at a rate γTC. APCs, A, are produced at a constant rate λ and die at a rate $\delta_1 A$. They take up antigen and become activated at a rate αAT. The parameter α summarizes several processes: the rate at which antigen is released from the cells, T, and the rate at which this antigen is taken up by APCs and processed for display and cross-presentation. Loaded APCs, A^*, are lost at a rate $\delta_2 A^*$. This corresponds either to death of the loaded APC, or to loss of the antigen-MHC complexes on the APC. Upon cross-presentation, CTL expand at a rate $\eta A^* C/(\epsilon C+1)$. The saturation term, $\varepsilon C+1$, has been included to account for the limited expansion of CTL in the presence of strong cross-stimulation [De Boer and Perelson (1995)]. The activated and expanding population of CTL can kill the infected cells upon direct presentation. In addition, it is assumed that direct presentation can result in removal of CTL at a rate qTC. This can be brought about, for example, by antigen-induced cell death, or over-differentiation into effectors followed by death. Finally, CTL die at a rate μC.

Thus a central assumption of the model is that cross-presentation can induce CTL expansion, while direct presentation does not have that effect; instead it can result in the decline of the CTL population. This assumption implies that the magnitude of cross-presentation relative to direct presentation could be a decisive factor which determines the outcome of a CTL response: activation or tolerance. In the model, the ratio of cross-presentation to direct presentation is given by cA^*/qT.

We assume that $r > a$. That is, the rate of increase of the target cells, T, is greater than their death rate. This ensures that this population of cells can grow and remain present. If this is fulfilled, the system can converge to a number of different equilibria (Figure 11.2). The expressions for the equilibria will not be written out here since most of them involve complicated expressions.

(1) The CTL response fails to expand, i.e. $C = 0$. The population of target cells grows to a high equilibrium level, unchecked by the CTL. The populations of unloaded and loaded APCs, A and A^*, also equilibrate.
(2) The CTL response expands, i.e. $C > 0$. In this case, the system can converge to one of two different outcomes. (a) The number of

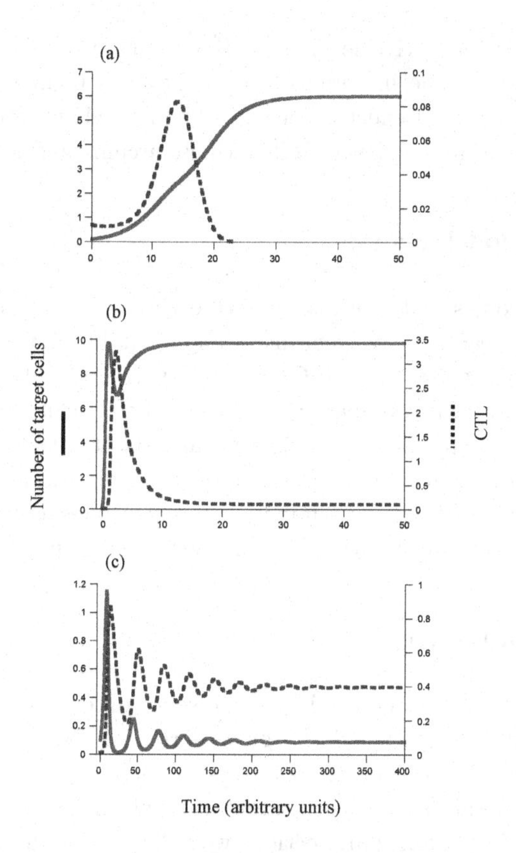

Fig. 11.2 Different outcomes of the model shown as time series. (a) Tolerance; CTL go extinct. (b) Tolerance outcome where CTL do not go extinct but are maintained at very low levels. (c) Immunity outcome. Parameters were chosen as follows: $r=0.5$; $k=10$; $d=0.1$; $\gamma=1$; $\lambda=1$; $\delta_1=0.1$; $\delta_2=1.5$; $\eta=2$; $\varepsilon=1$; $q=0.5$; $\mu=0.1$. $\alpha=0.2$ for (i) and $\alpha=0.1$ for (ii). For (iii) $\alpha=0.05$; $r=10$; $\eta=10$.

CTL is low and the number of target cells is high. This outcome is qualitatively similar to (i), because the CTL population does not fully expand, and the population of target remains high. (*b*) The number of CTL is high and the number of target cells is low. This can be considered the immune control equilibrium. If the population of target cells is reduced to very low levels, this can be considered equivalent to extinction (number of cells below one).

These equilibria therefore fall into two basic categories: (*a*) Tolerance; this is described by two equilibria. Either the immune response goes extinct, or it exists at low and ineffective levels. (*b*) Reactivity; this is described by only one equilibrium. The immune response expands to higher levels and exerts significant levels of effector activity. The following sections will examine which outcomes are achieved under which circumstances.

11.3 Method of model analysis

The equilibrium outcomes of the model describe the states to which the system can converge: reactivity or tolerance. These equilibria are roots of polynomials of degree larger than two. Consequently, the stability analysis of the equilibria was performed numerically, and will be presented as bifurcation plots below. The numerical analysis was carried out with a program called *Content.* It allows tracking the position of the equilibria as we vary parameters of the system. It also determines their eigenvalues as a function of parameters, thus giving full information on their stability properties.

11.4 Properties of the model

The two most important parameters in the present context are the rate of antigen uptake by APCs, α, and the growth rate of the target cells, r. This is because variation in these parameters can significantly influence the ratio of cross-presentation to direct presentation which is the subject of investigation. Hence, in the following sections we will examine the behavior of the model in dependence of these two parameters.

The rate of antigen uptake by APCs. The rate of antigen uptake by APCs comprises two processes: (i) the degree to which the antigen is made available for uptake; this can be determined for example by the amount of antigen released from the target cell, or the amount of apoptosis going on [Albert *et al.* (1998)]. (ii) The rate at which the APCs take up the available antigen and process it for presentation. As the rate of antigen uptake by APCs, α, decreases, the ratio of cross-presentation to direct presentation decreases (Figure 11.3a). When the value of α is high, the outcome is immunity. If the value of α is decreased and crosses a threshold, we enter a region of bistability (Figure 11.3a): both the immunity and the tolerance equilibria are stable. Which outcome is achieved depends on the initial conditions. If the value of α is further decreased and crosses another threshold, the immune control equilibrium loses stability. The only stable

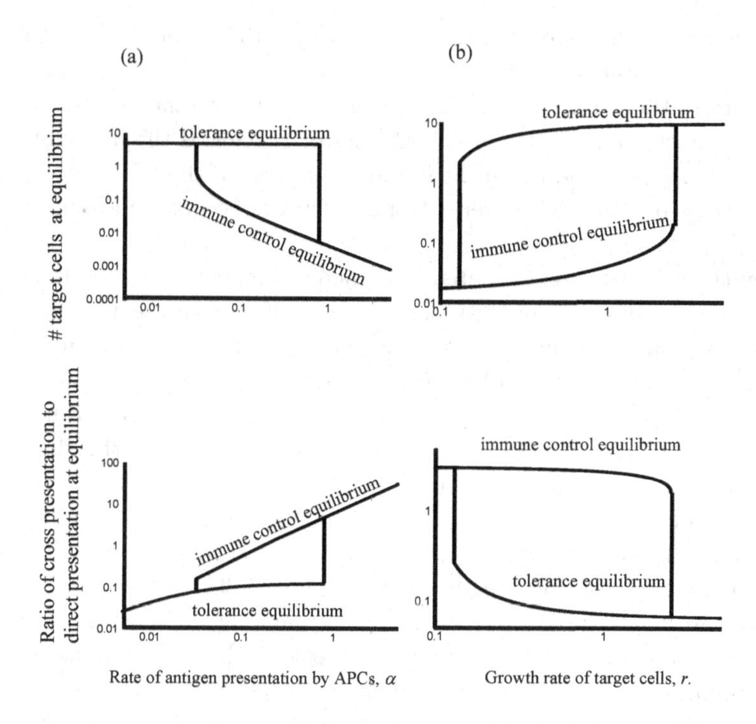

Fig. 11.3 Bifurcation diagram showing the outcome of the model as a function of (a) the rate of antigen presentation by APCs, α, and (b) the growth rate of target cells, r. Virus load and the ratio of cross-presentation to direct presentation at equilibrium are shown. Parameters were chosen as follows: $r=0.5$; $k=10$; $d=0.1$; $\gamma=1$; $\lambda=1$; $\delta_1=0.1$; $\alpha=0.5$; $\delta_2=1.5$; $\eta=2$; $\varepsilon=1$; $q=0.5$; $\mu=0.1$.

outcome is tolerance (Figure 11.3a).

In the region of bistability, the dependence on initial conditions is as follows. Convergence to the immune control equilibrium is promoted by low initial numbers of target cells, high initial numbers of presenting APCs, and high initial numbers of CTL. This is because under these initial conditions, the dynamics start out with a high ratio of cross-presentation to direct presentation and this promotes the expansion of the CTL. On the other hand, if the initial number of target cells is high and the initial number of presenting APCs and CTL is low, then the initial ratio of cross-presentation to direct presentation is low and this promotes tolerance. There are some slight variations to this general picture. As they do not alter the basic results, however, the reader is referred to [Wodarz and Jansen (2001)] for details.

In summary, as the rate of antigen uptake by APCs is decreased, the ratio of cross-presentation to direct presentation decreases, and this shifts the dynamics of the CTL response in the direction of tolerance. This can include a parameter region in which both the tolerance and the immunity outcome are stable, depending on the initial conditions. If the CTL responsiveness to cross-presentation is very strong, tolerance becomes an unlikely outcome.

The growth rate of target cells. An increase in the growth rate of target cells, r, results in a decrease in the ratio of cross-presentation to direct presentation in the model. Hence an increase in the growth rate of target cells shifts the dynamics of the CTL from a responsive state towards tolerance. The dependence of the dynamics on the parameter r is shown in Figure 11.3b. The growth rate of target cells needs to lie above a threshold to enable the CTL to potentially react. This is because for very low values of r, the number of target cells is very low, not sufficient to trigger immunity. If the growth rate of target cells is sufficiently high to potentially induce immunity, we observe the following behavior (Figure 11.3b). If the value of r lies below a threshold, the only stable outcome is immunity. If the value of r is increased and crosses a threshold, we enter a region of bistability. That is, both the immunity and the tolerance outcomes are possible, depending on the initial conditions. The dependence on the initial conditions is the same as explained in the last section. If the value of r is further increased and crosses another threshold, the immunity equilibrium loses stability and the only possible outcome is tolerance. Again, there are some slight variations to this general picture. As they do not alter the basic results, however, the reader is referred to [Wodarz and Jansen (2001)] for details.

In summary, an increase in the growth rate of target cells has a similar effect as a decrease in the rate of antigen uptake by APCs: the ratio of cross-presentation to direct presentation becomes smaller, and the outcome of the dynamics is driven from immunity towards tolerance. Again, this includes a parameter region where both the immunity and tolerance outcomes are stable and where the outcome depends on the initial conditions. The higher the overall responsiveness of the CTL to cross-stimulation, the less likely it is that a high growth rate of target cells can induce tolerance.

11.5 Immunity versus tolerance

The models have investigated the topic of CTL regulation from a dynamical point of view. We showed that the immune system can switch between two states: tolerance and activation. Which state is reached need not depend on the presence or absence of signals, but on the relative magnitude of cross-presentation to direct presentation. This shows that regulation can be accomplished without signals but in response to a continuously varying parameter. Thus, the regulation of CTL responses could be implicit in the dynamics. This relies on the assumption that there is a difference in the effect of cross-presentation and direct presentation. The mathematical model assumes that while cross-presentation results in CTL expansion, direct presentation results in lysis followed by removal of the CTL. Some mechanisms described in the literature support this notion. The simplest mechanism resulting in CTL removal could be antigen-induced cell death [Baumann *et al.* (2002); Budd (2001); Hildeman *et al.* (2002)]. That is, exposure to large amounts of antigen by direct presentation can trigger apoptosis in the T cells. Another mechanisms could be that exposure to direct presentation of antigen on the target cells induces the generation of short lived effectors which are destined for death [Guilloux *et al.* (2001)]. Since CTL effectors are thought to die shortly after killing target cells, exposure to large amounts of direct presentation can result in over-differentiation and an overall loss of CTL.

With the assumptions explained above, we find a very simple rule that determines whether CTL responses expand and react, or whether they remain silent and tolerant. CTL expansion and immunity is promoted if the ratio of cross-presentation to direct presentation is relatively high. This is because the amount of CTL expansion upon cross-presentation outweighs the degree of CTL loss upon direct presentation. On the other hand, tolerance is promoted if the ratio cross-presentation to direct presentation is relatively low. This is because the amount of CTL loss upon direct presentation outweighs the amount of CTL expansion upon cross-presentation.

For self antigen displayed on cells of the body, the ratio of cross-presentation to direct presentation is normally low. This is because these cells do not die at a high enough rate or release the antigen at a high enough rate for the amount of cross presentation to be strong. On the other hand, large amounts of this antigen can be available on the surface of the cells expressing them (direct presentation). In terms of our model, this situation can best be described by a low value of α. Hence, in our model,

CTL responses are not predicted to become established against self antigens. Instead, the outcome is tolerance. In addition, the initial conditions favor tolerance in this scenario. When immune cells with specificity for self are created and try to react, the number of these immune cells is very low and the number of target cells (tissue) is relatively high. This promotes failure of the CTL response to expand and to become established. On the other hand, with infectious agents, antigen is abundantly available. For example, virus particles are released from infected cells, ready to be taken up by APCs for cross-presentation. Therefore, the immune responses react and become fully established. In contrast, tumors may fail to induce CTL responses because tumor antigens are largely displayed on the surface of the tumor cells, but relatively little tumor antigen is made available for uptake by dendritic cells and hence for cross-presentation. The relative amount of cross-presentation, however, is influenced by parameters such as the growth rate of the target cells, and we observe a parameter region where the outcome of the CTL dynamics can depend on the initial conditions. Since cancer cells continuously evolve towards less inhibited growth, these results have implications for the role of CTL in tumor progression and cancer therapy. This is explored in the following sections.

11.6 Cancer initiation

A tumor cell is characterized by mutations which enable it to escape growth control mechanisms which keep healthy cells in check. According to the model, the generation of a tumor cell can lead to three different scenarios (Figure 11.4):

(i) A CTL response is induced which clears the cancer. (ii) A CTL response develops which is weaker; it controls the cancer at low levels, but does not eradicate it. (iii) A CTL response fails to develop; tolerance is achieved and the cancer can grow uncontrolled. Which outcome is attained depends on the characteristics of the cancer cells. In particular it depends on how fast the cancer cells can grow (r in the model), and how resistant they are against death and apoptosis. Cell death, and in particular apoptosis, is thought to increase the amount of cross-presentation [Albert *et al.* (1998)]. Resistance to apoptosis thus corresponds to a reduction in the parameter α in the model. Three parameter regions can be distinguished (Figure 11.5).

(i) If the cancer cells replicate slowly and/or still retain the ability to

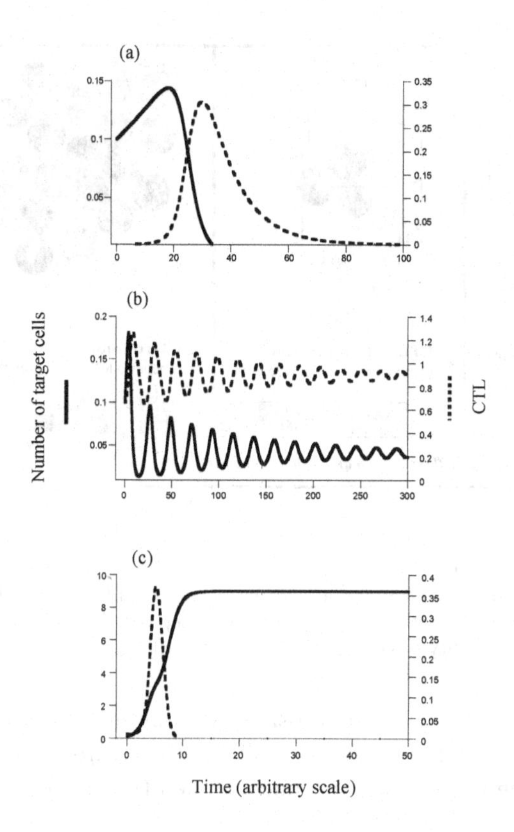

Fig. 11.4 Time series plots showing the different possible outcomes when a tumor is generated. (a) Clearance. (b) Immune control but failure to clear the target cells. (c) Tolerance. Parameters were chosen as follows: $k=10$; $d=0.1$; $\gamma=1$; $\lambda=1$; $\delta_1=0.1$; $\alpha=0.5$; $\delta_2=1.5$; $\eta=2$; $\varepsilon=1$; $q=0.5$; $\mu=0.1$. (a) $r=0.13$. (b, c) $r=1$. The difference between graphs (b) and (c) lies in the initial number of CTL, z.

undergo apoptosis, the cancer will be cleared, because strong CTL responses are induced. (ii) If the cancer cells replicate faster and/or the degree of apoptosis is weaker, the ratio of cross-presentation to direct presentation is reduced. This can shift the dynamics into the bistable parameter region. That is, the outcome depends on the initial conditions. If the initial size of the tumor is relatively small, it is likely that CTL responses will be develop successfully. This will result in control of the tumor. Because the response is less efficient, however, clearance is not likely. If the size of the tumor is already relatively large when the CTL response is activated, the likely outcome is tolerance. (iii) If the growth rate of the tumor cells is

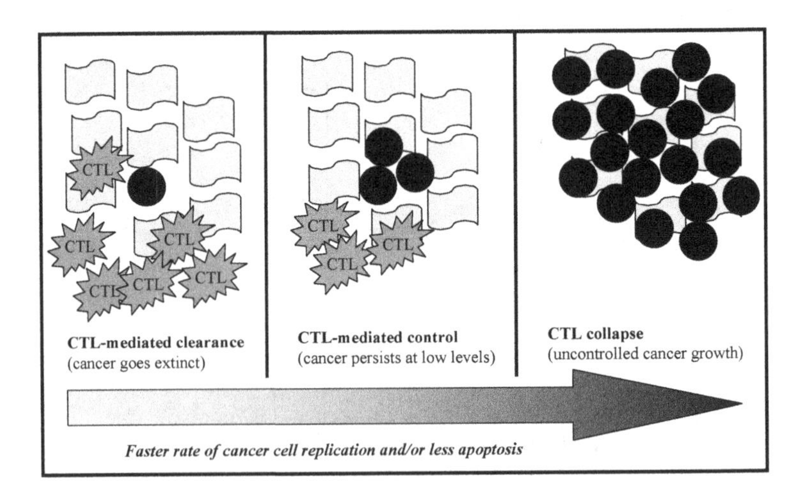

Fig. 11.5 Three parameter regions of the model: CTL mediated clearance, CTL-mediated control with tumor persistence at low levels, and CTL collapse leading to uncontrolled tumor growth. Which outcome is observed depends on the rate of cancer cell replication and on the ability of cells to undergo apoptosis.

still higher and/or the degree of apoptosis is still lower, then the ratio of cross-presentation to direct presentation falls below a threshold; now the only possible outcome is tolerance and uncontrolled tumor growth.

11.7 Tumor dormancy, evolution, and progression

Here, we investigate in more detail the scenario where the growth rate of the tumor is intermediate, and both the tolerance and the CTL control outcomes are possible, depending on the initial conditions. Assume the CTL control equilibrium is attained because the initial tumor size is small. The number of tumor cells is kept at low levels, but the tumor is unlikely to be cleared because in this bistable parameter region the ratio of cross-presentation to direct presentation is already reduced. If the tumor persists at low levels, the cells can keep evolving over time. They can evolve, through selection and accumulation of mutations, either towards a higher growth rate, r, or towards a reduced rate of apoptosis which leads to reduced levels of antigen uptake by dendritic cells, α. Both cases result in similar evolutionary dynamics. This is illustrated in Figure 11.6 assuming that the cancer cells evolve towards faster growth rates (higher values of

r).

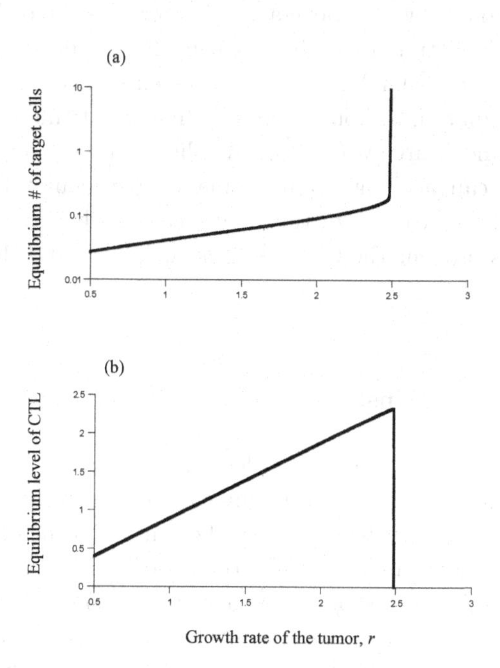

Fig. 11.6 Equilibrium tumor load (a) and the number of tumor specific CTL (b) as a function of the growth rate of tumor cells, r. This graph can by interpreted to show the effect of tumor evolution towards faster growth rates over time. As evolution increase the value of r over time, the tumor population and the CTL attain a new equilibrium. Parameters were chosen as follows: r=*0.5*; k=*10*; d=*0.1*; γ =*1*; λ=*1*; δ_1=*0.1*; α=*0.5*; δ_2=*1.5*; η =*2*; ε=*1*; q=*0.5*; μ =*0.1*.

An increase in the growth rate of tumor cells does not lead to a significant increase in tumor load. At the same time, it results in an increase in the number of tumor-specific CTL. The reason is that a faster growth rate of tumor cells stimulates more CTL which counter this growth and keep the number of tumor cells at low levels. When the growth rate of the tumor cells evolves beyond a threshold, the equilibrium describing CTL-mediated control of the cancer becomes unstable. Consequently, the CTL response collapses and the tumor can grow to high levels.

The dynamics of tumor growth and progression can include a phase called "dormancy". During this phase the tumor size remains steady at a low level over a prolonged period of time before breaking out of dormancy and progressing further. Several mechanisms could account for this phenomenon. The limitation of blood supply, or inhibition of angiogenesis, can prevent a tumor from growing above a certain size threshold [Folkman (1995b)]. When angiogenic tumor cell lines evolve, the cancer can progress further. Other mechanisms that have been suggested to account for dormancy are immune mediated, although a precise nature of this regulation remains elusive [Uhr and Marches (2001)]. As shown in this section, the model presented here can account for a dormancy phase in tumor progression. If the overall growth rate of the cancer cells evolves beyond this threshold, dormancy is broken: the CTL response collapses and the tumor progresses.

11.8 Immunotherapy against cancers

Assuming that the CTL response has failed and the cancer can grow unchecked, we investigate how immunotherapy can be used to restore CTL mediated control or to eradicate the tumor. In the context of the model, the aim of immunotherapy should be to increase the ratio of cross-presentation to direct presentation. The most straightforward way to do this is dendritic cell vaccination. In the model, this corresponds to an increase in the number of activated and presenting dendritic cells, A^*. We have to distinguish between two scenarios: (i) The tumor cells have evolved sufficiently so that the CTL control equilibrium is not stable anymore, and the only stable outcome is tolerance. (ii) The tumor has evolved and progressed less; the equilibrium describing CTL mediated control is still stable.

First we consider the situation where the tumor has progressed far enough so that the CTL control equilibrium is not stable anymore. Upon dendritic cell vaccination, tolerance is temporarily broken (Figure 11.7). That is, the CTL expand and reduce the tumor cell population. This CTL expansion is, however, not sustained and tumor growth relapses (Figure 11.7). The reason is as follows. Upon dendritic cell vaccination, the ratio of cross-presentation to direct presentation is increased sufficiently, enabling the CTL to expand. However, this boost in the level of cross-presentation subsequently declines, allowing the tumor to get the upper hand and regrow. The model suggests, however, that the tumor can be eradicated if the level of cross-presentation is continuously maintained at high levels. This can be achieved by repeated vaccination events (Figure 11.7). The next vaccination event has to occur before the level of cross-presentation has significantly declined. This will drive tumor load below a threshold level

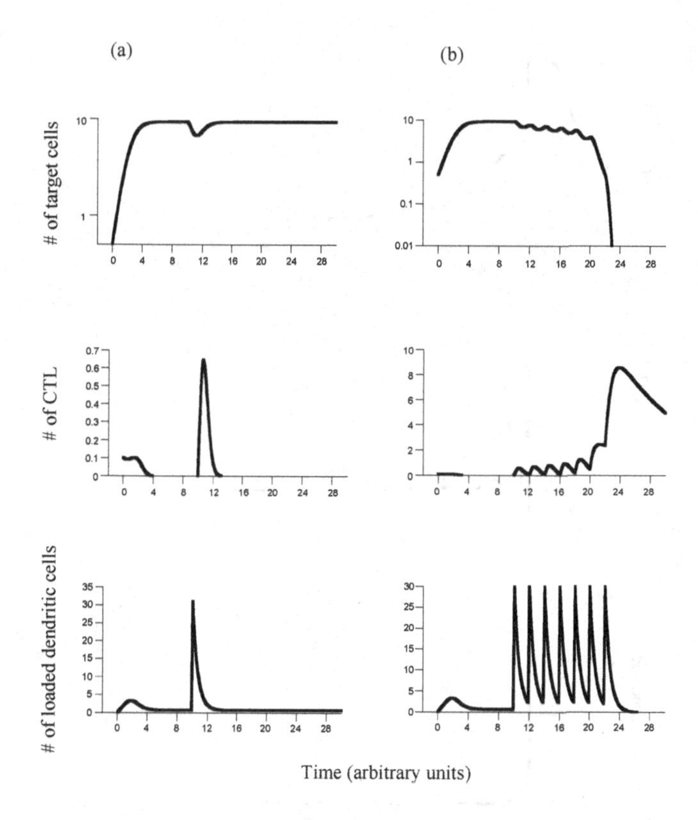

Fig. 11.7 Effect of dendritic cell vaccination on tumor dynamics assuming that the growth rate of the tumor has evolved to high values, where only the tolerance outcome is stable. (a) A single vaccination event induces a temporary reduction in tumor load, followed by a relapse. (b) Repeated vaccination events can drive the tumor load below a threshold which corresponds to extinction in practical terms. Parameters were chosen as follows: r=*1.5;* k=*10;* d=*0.1;* γ =*1;* λ=*1;* δ_1=*0.01;* α=*0.5;* δ_2=*1.5;* η =*0.5;* ε=*1;* q=*0.5;* μ =*0.1.*

which practically corresponds to extinction (Figure 11.7).

Next, we consider the more benign scenario in which the tumor has not progressed that far and the CTL control equilibrium is still stable. Now a single vaccination event can shift the dynamics from the tolerance outcome to the CTL control outcome (Figure 11.8). The reason is that an elevation in the number of presenting dendritic cells shifts the system into a space where the trajectories lead to CTL control and not to tolerance. This is likely to be achieved if the size of the tumor is not very large. The larger the size of the tumor, the stronger the vaccination has to be (higher A^*) in

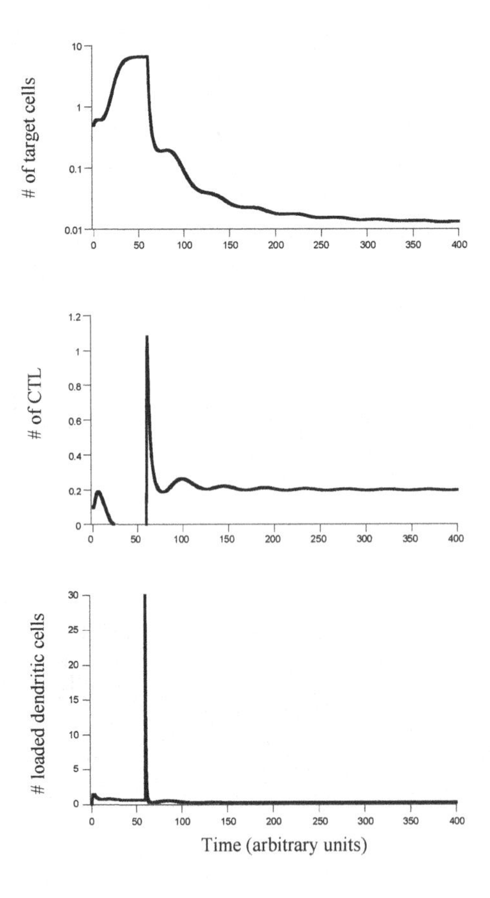

Fig. 11.8 Effect of dendritic cell vaccination on tumor dynamics assuming that the growth rate has not yet progressed beyond a threshold, so that we are in the bistable parameter region of the model. A single vaccination event can induce immunity which can control the tumor at low levels. Parameters were chosen as follows: $r=0.3$; $k=10$; $d=0.1$; $\gamma=1$; $\lambda=1$; $\delta_1=0.01$; $\alpha=0.5$; $\delta_2=1.5$; $\eta=0.5$; $\varepsilon=1$; $q=0.5$; $\mu=0.1$.

order to achieve success. If the tumor size is very large, then an elevated level of dendritic cells cannot shift the ratio of cross-presentation to direct presentation sufficiently to induce sustained immunity. A combination of vaccination and chemotherapy can, however, result in success. This is because chemotherapy reduces the size of the tumor and also induces death of tumor cells. Both factors contribute to a higher ratio of cross-presentation to direct presentation and to induction of immunity. Once a sustained CTL response has been induced, tumor cells are kept at low levels. However, the cancer is unlikely to be eradicated. Consequently, it can evolve over time. Thus, induction of CTL mediated control in the model is likely to result in a temporary phase of tumor dormancy. This phase is again broken after the overall growth rate of the tumor has evolved beyond the threshold at which the CTL control outcome becomes unstable.

These considerations result in the following suggestions. Dendritic cell vaccination should be administered repeatedly until the last tumor cell has been eradicated. If the tumor has already progressed relatively far, this is the only way to prevent immediate relapse of the cancer. If the tumor is less progressed, temporary tumor dormancy can be achieved by a single vaccination event. Tumor persistence and evolution will, however, break this dormancy phase, resulting in renewed cancer growth after a certain period of time. Thus, in this case, repeated vaccination is also advisable in order to keep the level of cross-presentation above a threshold and to avoid tumor persistence. In all cases, the model suggests that a combination of immunotherapy with conventional therapy is beneficial because conventional therapy can reduce the growth rate of the tumor. If conventional therapy increases the chances of developing immunological control of the tumor, conventional therapy would have to be applied only temporarily which would have significant clinical benefits.

Chapter 12

Therapeutic approaches: viruses as anti-tumor weapons

In the final chapter of this book we will discuss some treatment strategies against cancer. We start with a brief introductory section which summarizes some major treatment options. We then focus on an experimental gene therapeutic approach and show how mathematical analysis can be useful for identifying optimal treatment schedules.

The basic principle which underlies the treatment of cancer is the specific attack of cancer cells, without harming healthy cells. This is a most difficult challenge. The reason is that cancer cells are derived from healthy cells. While there are certainly many characteristics which are specific to cancer cells and which are not present in healthy cells, it is unclear how to exploit these differences therapeutically. Several therapy methods have already been mentioned or discussed in some detail throughout this book in different contexts. These include chemotherapy, immunotherapy, and the use of angiogenesis inhibitors. In the following we will briefly summarize common treatment options as well as some new and promising approaches.

As long as the tumor is localized and has not yet spread by metastasis, the best strategy is to remove it by surgery. This cures the patient. If the tumor has already spread to other tissues, however, surgery does not provide the patient with a significant benefit because the cancer cannot be eradicated anymore. In this case, the most common treatment strategy is chemotherapy, or radiation therapy if treatment is applied locally. These approaches work by damaging the DNA of cells. The aim is that this damage kills the cells. The rational for specificity is that chemotherapeutic agents are taken up by dividing cells. The common wisdom is as follows. Because cancer cells divide significantly more often than healthy tissue cells, chemotherapy is supposed to attack cancer cells preferentially. Tissue cells which do, however, divide relatively often, suffer side effects. Side effects are particularly common in epithelial tissue (hair, skin, mouth, digestive tract, etc.). It is unclear how chemotherapy works. Studies have attempted to find specific genes which correlate with success and failure. Some drugs appear to result in cell death by triggering apoptosis. Other drugs trigger cellular arrest and senescence. More work will be needed to develop genetic profiles which correlate with treatment success.

Recently, the use of targeted small molecules has caught a lot of attention as a complement or alternative to chemotherapy [Guillemard and Saragovi (2004); Smith *et al.* (2004)]. Research has identified molecular mechanisms and pathways which are responsible for cancerous growth. The aim is to use specific drugs to target these mechanisms and thereby remove the cancerous cells. There has already been some success with this kind of drugs. A Novartis drug called Gleevec has shown very promising results in the context of chronic myelogenous leukemia and a rare gastric cancer [Hasan *et al.* (2003); John *et al.* (2004)]. However, major problems remain, especially with respect to the emergence of drug resistance which leads to treatment failure [Yee and Keating (2003)]. In some cases, combining such targeted therapy with chemotherapy has been shown to have a beneficial impact on patients [Daley (2003)].

Another interesting therapy approach is still experimental. It serves as a nice example of how mathematical models can be used to help identify under which circumstances treatment success might be achieved. The treatment involves the use of so–called *oncolytic viruses* [Kirn *et al.* (1998); Kirn and McCormick (1996); McCormick (2003)]. These are viruses which have been engineered to replicate in cancer cells; healthy cells are not susceptible. They infect, reproduce, kill, and spread to further cancer cells. They act in a similar way as predators and biological control agents in agriculture. Predatory insects are used in order to destroy and control populations of pest insects which might be resistant to chemical pest control agents.

Several viruses have been altered to selectively infect cancer cells. Examples are HSV-1, NDV, and adenoviruses [Kirn and McCormick (1996)]. A specific example that has drawn attention recently is ONYX-015, an attenuated adenovirus which selectively infects tumor cells with a defect in p53 [Dix *et al.* (2000); Hall *et al.* (1998); Heise *et al.* (1999a); Heise *et al.* (1999b); Kirn and McCormick (1996); Oliff *et al.* (1996); Rogulski *et al.* (2000)]. This virus has been shown to have significant anti-tumor activity and has proven to be relatively effective at reducing or eliminating tumors in clinical trials in the context of head and neck cancer [Ganly *et al.* (2000); Khuri *et al.* (2000); Kirn *et al.* (1998)]. Yet challenges remain. In particular, it is unclear which virus characteristics are most optimal for therapeutic purposes. Viruses have been altered in a variety of ways by targeted mutations, but it is not clear what types of mutants have to be produced in order to achieve extinction of the cancer. Viruses can be altered with respect to their rate of infection, rate of replication, or the rate at which they kill cancer cells. Some studies have introduced "explosive" genes which the virus can deliver to the cancer cells and which will kill the cells instantly.

If tumor eradication does not occur, the outcome can be the persistence of both the tumor and the virus infection, and this would be detrimental for patients. Persistence of both tumor and virus has been seen in experiments with a mouse model system by Harrison et al. [Harrison *et al.* (2001)]. The reason for the failure to eradicate the tumor despite ongoing viral replication was left open to speculation.

Mathematical models have been used to address this question. Taking into account the complex interactions between viruses, tumor cells, and immune responses, such models have identified conditions under which oncolytic virus therapy is most likely to result in successful clearance of cancer. This chapter discusses these insights. The models take into account a variety of mechanisms which can contribute to cancer elimination. On the most basic level, virus infection and the consequent virus-induced death of the cancer cell can be responsible for tumor eradication. On top of this, the immune system is expected to have an effect. In particular, cytotoxic T lymphocytes (CTL, reviewed in Chapter 11) are likely to be important. These immune responses can kill cells which display foreign or mutated proteins. They may act in two basic ways. They can recognize the virus presented on infected cells and kill virus-infected cells. Alternatively, the virus infection may promote the establishment of a CTL response against cancer proteins, again resulting in their death. Mathematical models for each scenario will be discussed in turn, and some practical implications will be presented towards the end of the chapter.

12.1 Virus-induced killing of tumor cells

This section investigates the basic dynamics between a growing tumor population and a replicating virus selective for the tumor cells. Various aspects of tumor growth and inhibition have been modeled in literature [Adam and Bellomo (1997); Gatenby (1996); Gatenby and Gawlinski (1996); Kirschner and Panetta (1998)] and explored in previous chapters. Here, we concentrate on a simple model, capturing the essential assumptions for analyzing virus-mediated therapy. The model contains two variables: uninfected tumor cells, x, and tumor cells infected by the virus, y. It is explained schematically in Figure 12.1.

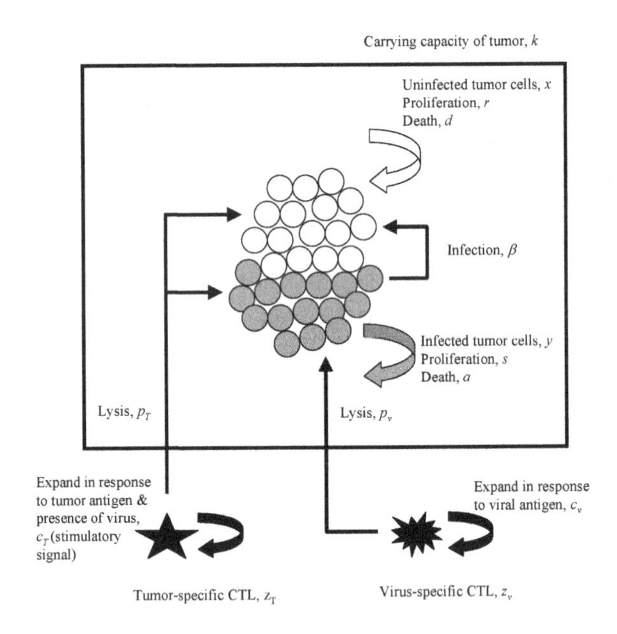

Fig. 12.1 Schematic representation of the models which are reviewed in this chapter.

The tumor cells grow in a logistic fashion at a rate r and die at a rate d. The maximum size or space the tumor is allowed to occupy is given by its carrying capacity k. The virus spreads to tumor cells at a rate β (this parameter can be viewed as summarizing the replication rate of the virus). Infected tumor cells are killed by the virus at a rate a and grow in a logistic fashion at a rate s. This assumes that division of infected tumor cells results in both daughter cells carrying the virus. This would certainly be the case with a virus that integrates into the tumor cell genome, but with a non-integrating virus, the chances of transmission upon cell division should be sufficiently high to justify this assumption. The model is given by the following set of ordinary differential equations [Wodarz (2001)].

$$\dot{x} = rx\left(1 - \frac{x+y}{k}\right) - dx - \beta xy,$$

$$\dot{y} = \beta xy + sy\left(1 - \frac{x+y}{k}\right) - ay.$$

In the absence of the virus the trivial equilibrium is attained and is given by E0: $x^{(0)} = k(r-d)/r$, $y^{(0)} = 0$.

The virus can establish an infection in the tumor cell population if $[\beta k(r-d) + sd]/r > a$. In this case, two types of outcomes are possible. The virus can either attain 100% prevalence in the tumor cell population (i.e. all tumor cells are infected), or it may only infect a fraction of the tumor cells (i.e. both uninfected and infected tumor cells are observed). Hundred percent virus prevalence is described by equilibrium E1:

$$x^{(1)} = 0, \quad y^{(1)} = k(s-a)/s.$$

Coexistence of infected and uninfected tumor cells is described by equilibrium E2:

$$x^{(2)} = \frac{\beta k\,(a-s) + ar - sd}{\beta\,(\beta k + r - s)}, \quad y^{(2)} = \frac{\beta k\,(r-d) + sd - ra}{\beta\,(\beta k + r - s)}.$$

The virus infects all tumor cells (equilibrium E1) if $a < s(d+\beta k)/(r+\beta k)$. Otherwise, equilibrium E2 is observed.

With this result in mind, how does viral cytotoxicity influence the size of the overall tumor? The tumor size is defined as the sum of infected and uninfected tumor cells, $x+y$, at equilibrium. Viral cytotoxicity has an opposing influence on tumor load depending on which equilibrium is attained (Figure 12.2). If all tumor cells are infected, then $x+y = k(s-a)/s$. An increase in viral cytotoxicity results in a reduction in tumor load (Figure 12.2). On the other hand, if not all tumor cells are infected, then $x+y = k(r-s+a-d)/(\beta k + r - s)$. Now, an increase in the viral cytotoxicity increases tumor load (Figure 12.2). The reason is that increased rates of tumor cell killing eliminate infected tumor cells before the virus had a chance to significantly spread. This in turn increases the pool of uninfected tumor cells and therefore the tumor load.

Hence, there is an optimal cytotoxicity, a_{opt}, corresponding to the minimum tumor size. This optimum is the degree of cytotoxicity at which the system jumps from the equilibrium describing 100% virus prevalence to the equilibrium where uninfected tumor cells are also present (Figure 12.2a).

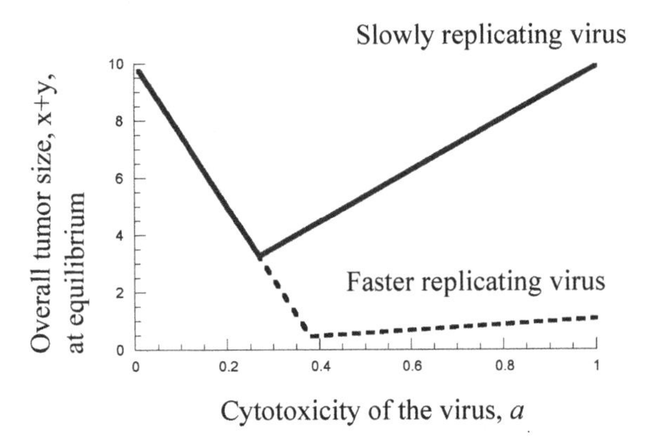

Fig. 12.2 Dependence of overall tumor load on the cytotoxicity of the virus. There is an optimal cytotoxicity at which tumor load is smallest. The faster the rate of virus replication, the higher the optimal level of cytotoxicity, and the smaller the minimum tumor load. Parameters were chosen as follows: $k=10$; $r=0.2$; $s=0.2$; $d=0.1$; for fast viral replication, $\beta=1$; for slow viral replication $\beta=0.1$.

The optimal viral cytotoxicity is thus given by $a_{opt} = s(d + \beta k)/(r + \beta k)$. At this optimal cytotoxicity the tumor size is reduced maximally and is given by $[x + y][min] = k(r - d)/(r + \beta k)$.

There are a number of points worth noting about this result. The minimum tumor size this therapy regime can achieve is most strongly determined by the replication rate of the virus, β (Figure 12.2). The higher the replication rate of the virus, the smaller the minimum size of the tumor. In order to achieve this minimum, the viral cytotoxicity must be around its optimum value. A major determinant of the optimal viral cytotoxicity is the rate of growth of uninfected and infected tumor cells (r and s respectively).

(1) If the infected tumor cells grow at a significantly slower rate relative to uninfected cells ($s << r$), the optimal cytotoxicity is low (Figure 12.3a). In the extreme case where the virus completely inhibits the ability of the tumor cell to divide, a non-cytotoxic virus is required to achieve optimal treatment results. More cytotoxic viruses result in tumor persistence (Figure 12.3a).

(2) On the other hand, if the growth rate of infected tumor cells is not significantly lower than that of uninfected tumor cells, an intermediate level of virus induced cell death is required to achieve minimum tumor

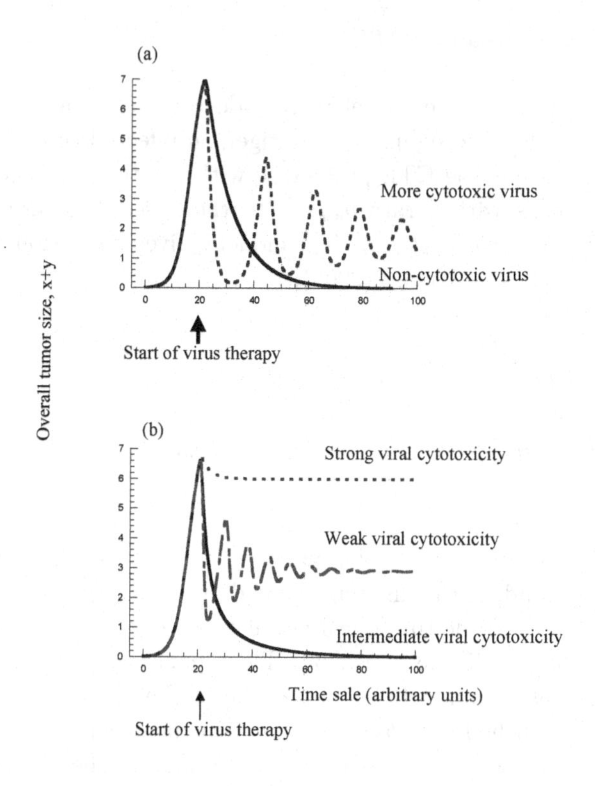

Fig. 12.3 Simulation of therapy using an oncolytic viruses in the absence of immunity. (a) The growth rate of infected tumor cells is significantly slower than that of uninfected tumor cells. A non-cytotoxic virus now results in tumor eradication. A more cytotoxic virus results in tumor persistence. Parameters were chosen as follows: k=*10*; r=*0.5*; s=*0*; β=*1*; d=*0.1*; $a = 0.1$ for the non-cytotoxic virus, and $a = 0.5$ for the more cytotoxic virus. (b) The growth rate of infected tumor cells is not significantly reduced relative to that of uninfected cells. An intermediate level of cytotoxicity results in tumor eradication. Weaker or stronger levels of cytotoxicity result in tumor persistence. Parameters were chosen as follows: k=*10*; r=*0.5*; s=*0*; β=*1*; d=*0.1*; $a = 0.2$ for the weakly cytotoxic virus, $a = 0.55$ for intermediate cytotoxicity, and $a = 3$ for strong cytotoxicity.

size (Figure 12.3b). If viral cytotoxicity is too weak, the tumor persists. However, if the viral cytotoxicity is too high, the tumor also persists because infected cells die too fast for the virus to spread efficiently (Figure 12.3b). In general, the faster the replication rate of the virus, the higher the optimal level of cytotoxicity.

12.2 Effect of virus-specific CTL

This section expands the above model to include a population of virus-specific CTL, z_v. The CTL recognize viral antigen on infected tumor cells. Upon antigenic encounter, the CTL proliferate with a rate $c_v y z_v$ and kill the infected tumor cells with a rate $p_v y z_v$. In the absence of antigenic stimulation the CTL die with a rate $b z_v$. The model is given by the following set of differential equations [Wodarz (2001)].

$$\dot{x} = rx\left(1 - \frac{x+y}{k}\right) - dx - \beta xy,$$
$$\dot{y} = \beta xy + sy\left(1 - \frac{x+y}{k}\right) - ay - p_v y z_v,$$
$$\dot{z} = c_v y z_v - b z_v.$$

First, we define the conditions under which an anti-viral CTL response is established. This condition is different depending on whether the virus attains 100% prevalence in the tumor cell population in the absence of the CTL. The strength of the CTL response, or CTL responsiveness, is denoted by c_v. If the virus has attained 100% prevalence in the absence of CTL, the CTL become established $c_v > bs/[k(s-a)]$. On the other hand, if the virus is not 100% prevalent in the tumor cell population in the absence of CTL, the CTL invade if $c_v > b\beta(\beta k + r - s)/[r(\beta k - a) - d(\beta k - s)]$.

In the presence of the CTL, we again observe two basic equilibria: either 100% virus prevalence in the tumor cell population, or the coexistence of infected and uninfected tumor cells. Hundred percent virus prevalence in the tumor cell population is described by equilibrium E1:

$$x^{(1)} = 0, \quad y^{(1)} = b/c_v, \quad z_v^{(1)} = \frac{kc_v(s-a) - sb}{p_v k c_v}.$$

Coexistence of infected and uninfected cells is described by equilibrium E2:

$$x^{(2)} = \frac{r\left(kc_v - b\right) - k\left(c_v d + b\beta\right)}{rc_v}, \quad y^{(2)} = b/c_v,$$

$$z_v^{(2)} = \frac{\beta k\left(rc_v - b\beta - c_v d\right) - c_v\left(ar - sd\right) - b\beta\left(r - s\right)}{p_v c_v r}.$$

How do the CTL influence the outcome of treatment? We distinguish between two scenarios.

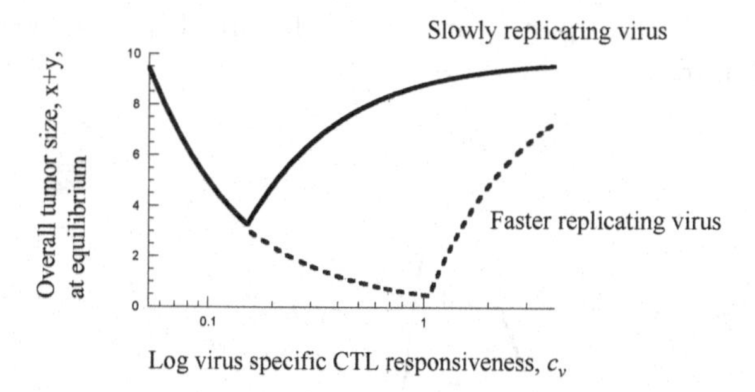

Fig. 12.4 Dependence of overall tumor load on the strength of the virus-specific CTL response. There is an optimal CTL responsiveness at which tumor load is smallest. The faster the rate of virus replication, the higher the optimal strength of the CTL response, and the smaller the minimum tumor load. Parameters were chosen as follows: $k=10$; $r=0.5$; $s=0.5$; $d=0.1$; $b=0.1$; $p=1$; $a=0.2$; for fast viral replication, $\beta=1$; for slow viral replication $\beta=0.1$.

(*i*) If the virus has established 100% prevalence in the tumor cell population in the absence of the CTL response, the presence of CTL can both be beneficial and detrimental to the patient (Figure 12.4). On one hand, the virus can remain 100% prevalent in the tumor in the presence of CTL. In this case, overall tumor size is given by $x + y = b/c_v$. At this equilibrium, an increase in the CTL responsiveness against the virus decreases the tumor size. On the other hand, if the CTL responsiveness crosses a threshold given by $c_v > b(\beta k + r)/[k(r - d)]$, the virus does not maintain 100% prevalence in the tumor cell population, and the overall tumor size is given by $x + y = k[c_v(r - d) - b\beta]/(c_v r)$. In this case, an increase in the CTL responsiveness to the virus increases tumor load and is detrimental to the patient (Figure 12.4). This is because the CTL response kills the virus faster than it can spread. Hence, the optimal CTL responsiveness is given by $c_{opt} = b(\beta k + r)/[k(r - d)]$. At this optimal CTL responsiveness, the tumor size is reduced maximally and is given by $[x + y][min] = k(r - d)/(r + \beta k)$. The faster the replication rate of the virus, the higher the optimal CTL responsiveness, and the lower the minimum size of the tumor that can be attained by therapy (Figure 12.4). Note that the minimum tumor size that can be achieved is the same as in the previous case where viral cytotoxicity alone was responsible for reducing the tumor. The effect of the CTL response is to modulate the overall death rate of infected cells with the aim of pushing it towards its optimum value. Figure 12.5 shows a simulation of therapy where an intermediate CTL responsiveness results in tumor remission, while a stronger CTL response can result in failure of therapy because virus spread is inhibited.

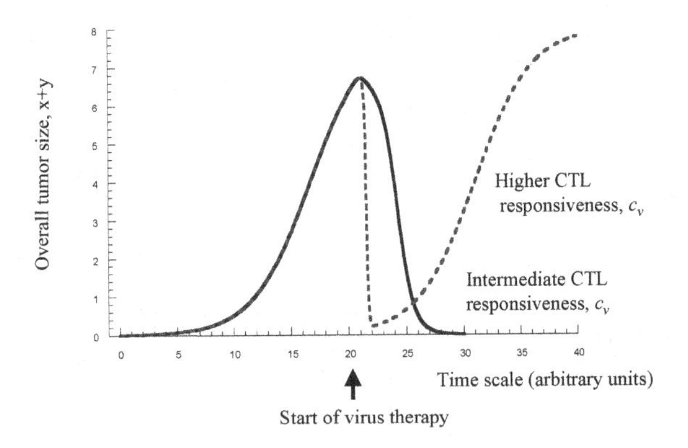

Fig. 12.5 Simulation of therapy using an oncolytic viruses in the presence of virus-specific lytic CTL. An intermediate CTL responsiveness results in tumor eradication, while a stronger CTL response results in tumor persistence. Parameters were chosen as follows: k=*10;* r=*0.5;* s=*0.5;* β=*0.1;* a=*0.2;* p=*1;* b=*0.1;* b=*0.1;* d=*0.1;* The intermediate CTL responsiveness is characterized by $c_v = 0.2625$, while the stronger CTL response is characterized by $c_v = 2$.

(ii) If the virus is not 100% prevalent already in the absence of the CTL response, a CTL-mediated increase in the death rate of infected cells can only be detrimental to the patient since it increases tumor load. The system converges to an equilibrium tumor size described by $x + y = k[c_v(r - d) - b\beta]/c_v r$.

12.3 Virus infection and the induction of tumor-specific CTL

Previous sections explored how virus infection and the virus-specific CTL response can influence tumor load. However, virus infection might not only induce a CTL response specific for viral antigen displayed on the surface of the tumor cells. In addition, active virus replication could induce a CTL response specific for tumor antigens [Fuchs and Matzinger (1996); Matzinger (1998)]. The reason is that virus replication could result in the release of substances and signals alerting and stimulating the immune system. This could be induced by tumor antigens being released and taken up by professional antigen presenting cells (APC), and/or by other signals released from the infected tumor cells. This is known as the danger signal hypothesis in immunology (and is discussed in more detail in Chapter 11). Normal tumor growth is thought not to evoke such signals, whereas the presence of viruses might evoke danger signals. Here, such a tumor specific CTL response is included in the model. It is assumed that the responsiveness of the tumor-specific CTL requires two signals: (i) the presence of the tumor antigen, and (ii) the presence of infected tumor cells providing immunostimulatory signals. In the following, the interactions between the tumor, the virus, and the tumor-specific CTL are investigated.

A model is constructed describing the interactions between the tumor population, the virus population, and a tumor-specific CTL response. It takes into account three variables. Uninfected tumor cells, x, infected tumor cells, y, and tumor-specific CTL, z_T. It is given by the following set of differential equations [Wodarz (2001)],

$$\begin{aligned}
\dot{x} &= rx\left(1 - \frac{x+y}{k}\right) - dx - \beta xy - p_T x z_T,\\
\dot{y} &= \beta xy + sy\left(1 - \frac{x+y}{k}\right) - ay - p_T y z_T,\\
\dot{z}_T &= c_T(x+y)z_T - bz_T.
\end{aligned}$$

The basic interactions between viral replication and tumor growth are identical to the models described above. The tumor-specific CTL expand in response to tumor antigen, which is displayed both on uninfected and infected cells ($x+y$), at a rate c_T. However, in accord with the danger signal hypothesis, it is assumed that the tumor-specific CTL response only has the potential to expand in the presence of the virus, y. In the model virus load correlates with the ability of the tumor-specific response to expand, since high levels of viral replication result in stronger stimulatory signals. The tumor-specific CTL kill both uninfected and infected tumor cells at a rate $p_T y z_T$.

If the virus has reached 100% prevalence in the absence of CTL, the tumor-specific CTL response becomes established if $c_T > bs^2/[k(a-s)]^2$. If infected and uninfected tumor cells coexist in the absence of CTL, the tumor-specific CTL response becomes established if

$$c_T > \frac{b\beta(s - r - \beta k)^2}{k[\beta k(r-d) - ra + sd](r - s + a - d)}.$$

In the presence of the tumor-specific CTL, the virus can again attain 100% prevalence in the tumor cell population, or we may observe the coexistence of infected and uninfected tumor cells. Hundred percent prevalence in the tumor population is described by equilibrium E1:

$$x^{(1)} = 0, \quad y^{(1)} = (b/c_T)^{1/2}, \quad z_T^{(1)} = \frac{k\,(s-a) - sy^{(1)}}{p_T k}.$$

Coexistence of infected and uninfected tumor cells is described by equilibrium E2:

$$y^{(2)} = \frac{b\beta}{ac_T}, \quad z_T^{(2)} = \frac{1}{p_T}\left[r\left(1 - \frac{x^{(2)} + y^{(2)}}{k}\right) - d - \beta y^{(2)}\right],$$

$$x^{(2)} = \frac{Q}{c_T k\,(\beta k + r - s)\,(r - d + a - s)},$$

where

$$Q = c_T rk\,[r - 2\,(d + s + a)] + c_T dk^2\,[d + 2\,(s - a)] + c_T k^2\left[s\,(s - 2a) + a^2\right] + b\beta k\,[2\,(s - r) - \beta k] - br^2 + bs\,(2r - s)\,.$$

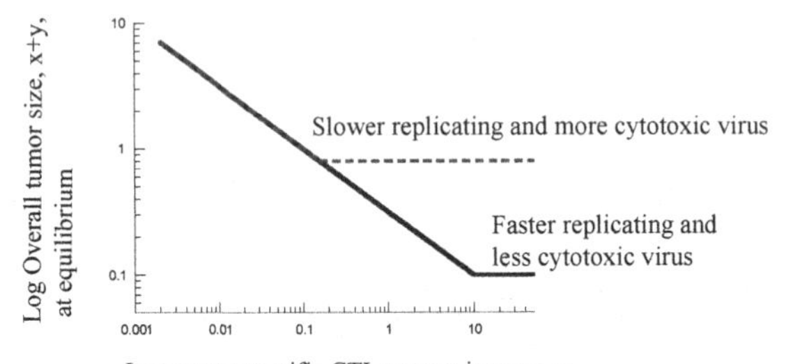

Fig. 12.6 Dependence of overall tumor load on the strength of the tumor-specific CTL response. The higher the strength of the tumor-specific CTL, the lower tumor load. If the strength of the tumor-specific CTL crosses a threshold, tumor load becomes independent of CTL parameters. The faster the rate of virus replication and the smaller the degree of viral cytotoxicity, the further the overall tumor load can be reduced. Parameters were chosen as follows: k=*10;* r=*0.5;* s=*0.5;* d=*0.1;* b=*0.1*. The fast replicating and weakly cytotoxic virus is characterized by β=*1* and $a = 0.2$. The slower replicating and more cytotoxic virus is characterized by β=*0.5* and $a = 0.5$.

We investigate how the responsiveness of the tumor-specific CTL, c_T, influences the size of the tumor, $x + y$. The presence of the tumor-specific CTL can have the following effects. If the virus achieves 100% prevalence in the tumor cell population, then $x + y = (b/c_T)^{1/2}$. Thus, an increase in the responsiveness of the tumor-specific CTL results in a decrease in tumor load (Figure 12.6). If $c_T > b(\beta k + r - s)^2/[k(r - s + a - d)]^2$, the virus is not 100% prevalent in the tumor cell population. This switch is thus promoted by a high responsiveness of the tumor-specific CTL relative to the replication rate of the virus (Figure 12.6). In this case, the size of the tumor is given by $x + y = k(r - s + a - d)/(\beta k + r - s)$. This is the minimum tumor size that can be achieved. Thus, if the CTL responsiveness against the tumor lies above a threshold, tumor load reaches its minimum (Figure 12.6). Note that it also becomes independent of the strength of the CTL. Hence, a CTL responsiveness that lies above this threshold is not detrimental to the patient. In this situation, tumor size is determined by the replication rate and the cytotoxicity of the virus (Figure 12.6). The higher the replication rate of the virus and the lower the degree of viral cytotoxicity, the smaller the tumor. The reason is that fast viral replication and low cytotoxicity result in higher virus load which in turn results in stronger signals to induce the tumor-specific CTL. Figure 12.7 shows a simulation of treatment underscoring this result.

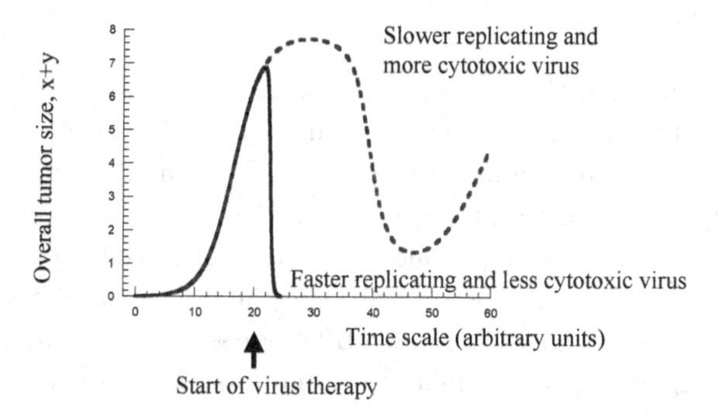

Fig. 12.7 Simulation of therapy using an oncolytic virus in order to stimulate a tumor-specific CTL response. If the virus replicates at a fast rate and is weakly cytotoxic, the level of immuno-stimulatory signals is high. Hence the tumor-specific response is strong and drives the tumor extinct. Parameters were chosen as follows: k=*10*; r=*0.5*; s=*0.5*; d=*0.1*; b=*0.1*; $c_T = 0.2$. The fast replicating and weakly cytotoxic virus is characterized by β=*0.5* and $a = 0.2$. The slower replicating and more cytotoxic virus is characterized by β=*0.1* and $a = 0.6$.

A note of caution: the model assumes that the production of immuno-stimulatory signals induced by the virus is proportional to the amount of viral replication. If cellular debris following virus-mediated destruction of cells also contributes to these signals, then the effect of viral cytotoxicity could be more complex. However, the exact nature and concept of the so–called danger signals is still controversial. The model takes into account the simple observation that presence of signals typical of viral replication can enhance immunity to tumors.

12.4 Interactions between virus- and tumor-specific CTL

In this section, the two types of CTL responses studied above are brought together. That is, both the virus- and the tumor specific CTL responses are taken into consideration. The model is explained schematically in Figure 12.1 and given by the following set of differential equations [Wodarz (2001)]:

$$
\begin{aligned}
\dot{x} &= rx\left(1-\frac{x+y}{k}\right) - dx - \beta xy - p_T x z_T,\\
\dot{y} &= \beta xy + sy\left(1-\frac{x+y}{k}\right) - ay - p_v y z_v - p_T y z_T,\\
\dot{z}_v &= c_v y z_v - b z_v,\\
\dot{z}_T &= c_T(x+y) z_T - b z_T.
\end{aligned}
$$

In this model the virus- and the tumor specific CTL responses are in competition with each other, because both can reduce tumor load and hence the strength of the stimulus required to induce CTL proliferation. In the following these competition dynamics are examined.

If the virus has reached 100% prevalence in the tumor cell population in the absence of CTL, then virus- and tumor specific CTL cannot coexist. If $c_v > (c_T b)^{1/2}$, then the virus-specific CTL response is established. On the other hand, if $c_v < (c_T b)^{1/2}$, then the tumor-specific CTL response becomes established.

If both infected and uninfected tumor cells are present in the absence of CTL, the situation is more complicated. Now, three outcomes are possible. Either the virus-specific response becomes established, or the tumor-specific response becomes established, or both responses can coexist. The virus-specific response persists if $c_v > kc_T(r-s+a-d)/(\beta k+r-s)$. The tumor-specific response persists if $c_T > c_v^2 r/\{k[c_v(r-d)-b\beta]\}$. Coexistence of both CTL responses is only observed if both of these conditions are fulfilled. This outcome is described by the following equilibrium expressions:

$$x^{(1)} = \frac{c_v^2 - bc_T}{c_v c_T}, \quad y^{(1)} = b/c_v,$$

$$z_T^{(1)} = \frac{1}{p_T}\left[r\left(1 - \frac{x^{(1)} + y^{(1)}}{k}\right) - d - \beta y^{(1)}\right],$$

$$z_v^{(1)} = \frac{1}{p_v}\left[\beta x^{(1)} + s\left(1 - \frac{x^{(1)} + y^{(1)}}{k}\right) - a - p_T z_T^{(1)}\right].$$

If both responses coexist, then the size of the tumor is given by $x + y = c_v/c_T$. Thus, a strong tumor-specific response, c_T, reduces tumor load. On the other hand, a strong virus-specific response, c_v, increases tumor load. The reason is that a strong virus-specific response results in low virus load and therefore in low stimulatory signals promoting the induction of tumor-specific immunity. Note that this last statement only applies to the parameter region where both types of CTL responses co-exist.

12.5 Treatment strategies

The above discussion has shown that the outcome of therapy depends on a complex balance between host and viral parameters. An important variable is the death rate of infected tumor cells. In order to achieve maximum reduction of the tumor, the death rate of the infected cells must be around its optimum, defined by the mathematical models. If the death rate of infected cells lies around its optimum, a fast replication rate of the virus and a slow growth rate of the tumor increase the chances of tumor eradication. The death rate of infected tumor cells can be influenced by a variety of factors: (*i*) Viral cytotoxicity alone kills tumor cells. *(ii)* A CTL response against the virus contributes to killing infected tumor cells. *(iii)* The virus helps eliciting a tumor-specific CTL response following the release of immuno-stimulatory signals.

The most straightforward way to use viruses as anti-cancer weapons is in the absence of immunity. If the cytotoxicity of the virus is around its optimum value, minimum tumor size is achieved. It is important to note that the highest rate of virus induced tumor cell killing does not necessarily contribute to the elimination of the tumor. The reason is that a very high rate of virus-induced cell death compromises the overall spread of the infection through the tumor. If a virus specific CTL response is induced, the best strategy would be to use a fast replicating and weakly cytotoxic virus. This is because the CTL will increase the death rate of infected cells. If the overall death rate of infected cells is too high, this is detrimental to the patient, since virus spread is prevented. In addition, a weakly cytotoxic and fast replicating virus may provide the strongest stimulatory signals for the establishment of tumor-specific immunity.

Because the model suggests that a fast growth rate of the tumor decreases the efficacy of treatment, success of therapy could be promoted by using a combination of virus therapy and conventional chemo- or radiotherapy. These suggestions are supported by recent experimental data [Freytag *et al.* (1998); Heise *et al.* (1997); Rogulski *et al.* (2000); You *et al.* (2000)]. A combination of treatment with the adenovirus ONYX-015 and chemotherapy or radiotherapy has been shown to be significantly more effective than treatment with either agent alone.

The principles of the mathematical modeling approaches presented here can help to improve treatment and to attain higher levels of success. In order to achieve this, however, more work is needed. Basic parameters of viruses and virus mutants need to be measured as a first step. Because the optimal death rate of infected tumor cells is crucial, it will be important to precisely measure the rate at which different viruses kill the tumor cells. Equally important is the quantification of the viral replication kinetics. Once such basic parameters have been measured, it is important to re-consider some model assumptions. The models discussed in this chapter are only a first approach to use computational methods for the analysis of oncolytic virus therapy, and the models will probably need to be revised and improved. For example, it is unclear whether and how the replication rate of the virus correlates with the rate of virus-induced cell killing. Many possibilities exist, and this is similar to the relationship between pathogen spread and "virulence" in an epidemiological context. Such more detailed information, based on experimental measurements, will be important to incorporate into the models in order to make more solid and reliable predictions.

12.6 Evaluating viruses in culture

A central result derived from the mathematical models is that success is promoted by using a virus which induces an optimal death rate of infected cells. Too high a rate of virus-induced cell death is detrimental and leads to the persistence of both tumor and virus, because overall virus spread is impaired. This gives rise to important insights for the methods used to evaluate potential viruses in culture [Wodarz (2003)].

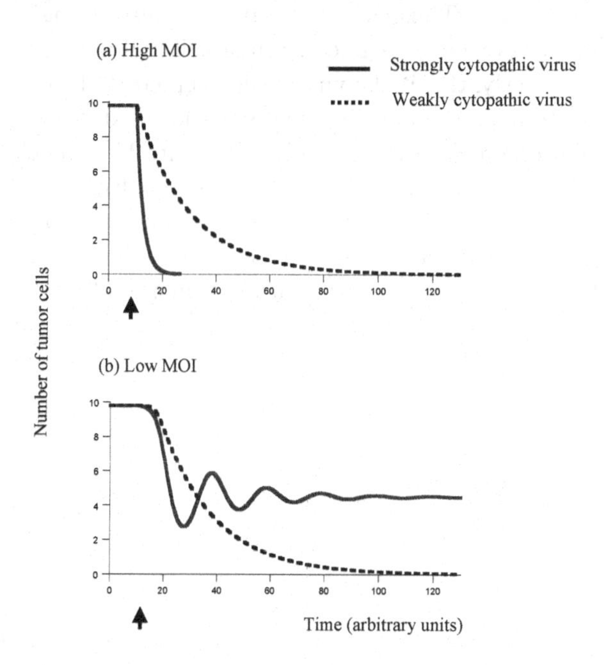

Fig. 12.8 Simulation showing the evaluation of potential replicating viruses in culture. A weakly and a strongly cytopathic virus are compared. Introduction of the virus is indicated by an arrow. (a) High multiplicity of infection. (b) Low multiplicity of infection. Parameters were chosen as follows: r=0.5; s=0; k=10; β=1.5; d=0.01; k=0.1; u=1. For the strongly cytopathic virus, $a = 0.4$. For the weakly cytopathic virus, $a = 0.04$. Virus inoculum was $y = 10$ for high MOI and $y = 0.01$ for low MOI.

In vitro experiments can be used to evaluate the potential efficiency with which the virus can eradicate a tumor. This is done by infecting a population of cancer cells with virus in a dish and monitoring the number of cancer cell over time. The models suggest that a low multiplicity of infection (MOI, i.e. the initial abundance of the virus relative to the tumor cells) is required to evaluate the virus. The reason is that *in vivo*, the replicating virus has to spread through the cancer cell population, and this has to be mimicked in culture. Using a high MOI can lead to misleading evaluations. These notions are illustrated in Figure 12.8 with computer simulations. This figure depicts the dynamics in culture for strongly and weakly cytopathic viruses, using different MOIs. Figure 12.6a shows the dynamics for a high MOI. In this simulation, the strongly cytopathic virus results in quick elimination of the tumor cells, while the weakly cytopathic virus is much less effective. Thus, if viruses are evaluated using a high MOI, the virus with the strongest degree of tumor cell killing receives the highest grades. Importantly, this is the virus which is predicted to be least efficient at reducing tumor load *in vivo.* The situation is different when viruses are evaluated in culture using a low MOI (Figure 12.6b). The less cytopathic virus results in elimination of tumor cells in culture, while the more cytopathic virus fails to eliminate tumor cells in culture. Therefore, the less cytopathic virus gets the better marks, and this is also the virus which is predicted to be more efficient at reducing tumor load *in vivo.*

Appendix A

Exact formula for total probability of double mutations

Here we calculate the probability of having a double mutant in a crypt, as a function of time. We focus on the progeny of one stem cell. The total number of cells in the crypt, $N(n)$, is

$$N(n) = \begin{cases} 2^n, & n < l \\ 2^l & n \geq l. \end{cases}$$

Note that $n < l$ represents the "development" stage of the crypt. After $n = l$, the size of the crypt remains constant. Each cell division may result in creating mutants. The probability of mutation is p_1, so that with probability $2p_1(1 - p_1)$ one mutant is created, and with probability p_1^2 two mutants are created. A mutated cell can acquire a second mutation with probability p_2.

Let us trace mutations in one clone. The number of mutants at time n (with one mutation) is a random variable, whose values go from 0 to 2^{l-1}. If a double mutant is created, we set this random variable to the value E (for "end"). This is an absorbing state because once a double mutant is creat|ed, it is assumed to stay in the system. We have a time-inhomogeneous Markov process for the variable j, $j \in \{0, 1, \ldots, 2^{l-1}, E\}$. The probability distribution for the variable j at time n is a row-vector $p^{(n)}$, whose evolution is given by

$$p^{(n)} = p^{(n-1)} M^{(n)}, \quad n = 1, 2, \ldots.$$

The transition matrices, $M^{(n)}$, can be easily written down. The first matrix corresponds to the asymmetric division of the SC, and is given by

$$M_{00}^{(1)} = 1 - p_1, \quad M_{01}^{(1)} = p_1, \quad M_{11}^{(1)} = 1 - p_2, \quad M_{1E}^{(1)} = p_2, \quad M_{EE}^{(1)} = 1,$$

with all the rest of the elements being zero. The matrices corresponding to symmetric divisions of DCs are given by the following:

$$M^{(n)}_{j,2j+k} = \binom{2^{n-1}-2j}{k} p_1^k (1-p_1)^{2^{n-1}-2j-k}(1-p_2)^{2j},$$

for $0 \leq k \leq 2^{n-1}-2j, \quad 0 \leq j \leq 2^{n-2}$, and

$$M^{(n)}_{j,E} = 1-(1-p_2)^{2j}, \quad 0 \leq j \leq 2^{n-2},$$

$$M^{(n)}_{E,E} = 1,$$

with all the rest of the elements being zero. Let us suppose that the SC is not mutated at time $n=0$. Let us consider one clone in isolation. The probability to have at least one double mutant by time n is given by

$$P_0(E;n) = \left[(1,0,\ldots)\prod_{s=1}^{n} M^{(s)}\right]_E, \quad n \leq l, \quad P_0(E;n) = P_0(E;l), \quad n > l;$$

here $[\ldots]_E$ stands for the Eth entry of a row-vector. Note that after l divisions starting from SC, the cumulative probability does not change anymore because the clone is removed. Similarly, if the SC is mutated at time 0, then the corresponding probability is

$$P_*(E;n) = \left[(0,1,\ldots)\prod_{s=1}^{n} M^{(s)}\right]_E, \quad n \leq l, \quad P_*(E;n) = P_*(E;l), \quad n > l.$$

This formula can be simplified to give

$$P_*(E;n) = \sum_{j=0}^{\mathcal{N}(n)-1} (1-(1-p_2)^{2^j})(1-p_2)^{2^j-1},$$

where the upper limit of summation is $\mathcal{N}(n) = n$ for $n \leq l$, and $\mathcal{N}(n) = l$ for $n > l$.

We would like to calculate the probability to have at least one double mutant in the system by time n. We consider a sequence of clones. There are n divisions of the SC. Let us denote the event of acquiring a mutation in the SC at time k, and no further mutations in the SC before time n, by $S_{k,n}$. We have

$$Prob(S_{k,n}) = (1-p_1)^{k-1} p_1 (1-p_2)^{n-k}.$$

Given $S_{k,n}$, the probability to have no double mutant by time n is

$$Prob(\bar{E};n|S_{k,n}) = \prod_{j=1}^{k}(1 - P_0(E;n-j+1)) \prod_{j=k+1}^{n}(1 - P_*(E;n-j+1)).$$

Each clone is originated at time j, so its "age" is $n-j+1$. Similarly, the probability to have no double mutant by time n, given that no SC mutations have occurred by time n, is

$$Prob(\bar{E};n|\text{no SC mut}) = \prod_{j=1}^{n}(1 - P_0(E;n-j+1)),$$

and the probability to have no SC mutations by time n is $(1-p_1)^n$. The total probability to have at least one double mutant by time n is given by

$$1 - Prob(E;n) = \sum_{k=1}^{n} Prob(\bar{E};n|S_{k,n})Prob(S_{k,n}) + Prob(\bar{E};n|\text{no SC mut})(1-p_1)^n. \quad \text{(A.1)}$$

This formula was compared with numerical simulations for the total probability of a double mutant, and with the approximate expression for $x^{ss} + x^{sd} + x^{dd}$ derived in Appendix A. The agreement with the numeric is perfect. The agreement with the approximation is very good, if the following consideration is taken into account. Note that the expressions for x^{ss}, x^{sd} and x^{dd} (formulas (5.5), (5.8) and (5.9)) were derived assuming that the size of the crypt is always N. On the other hand, in the simulations and in formula (A.1) we have a "development" phase where the size of the crypt is growing. The growth is exponential and thus during the first $l-2$ steps, the population size is significantly less than N, and thus mutation probabilities are negligible. This is why in order to get a perfect fit, the expressions for x^{ss}, x^{sd} and x^{dd} should be shifted by $l-2$ steps.

Bibliography

Adam, J. A. and Bellomo, N. (1997) *A survey of models for tumor-immune system dynamics.* Boston: Birkhauser.

Ahuja, N., Mohan, A. L., Li, Q., Stolker, J. M., Herman, J. G., Hamilton, S. R., Baylin, S. B. and Issa, J. P. (1997) Association between CpG island methylation and microsatellite instability in colorectal cancer. *Cancer Res* **57**, pp. 3370-4.

Alarcon, T., Byrne, H. M. and Maini, P. K. (2003) A cellular automaton model for tumour growth in inhomogeneous environment. *J Theor Biol* **225**, pp. 257-74.

Albert, M. L., Jegathesan, M. and Darnell, R. B. (2001) Dendritic cell maturation is required for the cross-tolerization of CD8+ T cells. *Nat Immunol* **2**, pp. 1010-7.

Albert, M. L., Sauter, B. and Bhardwaj, N. (1998) Dendritic cells acquire antigen from apoptotic cells and induce class I-restricted CTLs. *Nature* **392**, pp. 86-9.

Albertini, R. J., Nicklas, J. A., O'Neill, J. P. and Robison, S. H. (1990) In vivo somatic mutations in humans: measurement and analysis. *Annu Rev Genet* **24**, pp. 305-26.

Albertson, D. G., Collins, C., McCormick, F. and Gray, J. W. (2003) Chromosome aberrations in solid tumors. *Nat Genet* **34**, pp. 369-76.

Amon, A. (1999) The spindle checkpoint. *Curr Opin Genet Dev* **9**, pp. 69-75.

Andea, A. A., Wallis, T., Newman, L. A., Bouwman, D., Dey, J. and Visscher, D. W. (2002) Pathologic analysis of tumor size and lymph node status in multifocal/multicentric breast carcinoma. *Cancer* **94**, pp. 1383-90.

Anderson, A. R. and Chaplain, M. A. (1998) Continuous and discrete mathematical models of tumor-induced angiogenesis. *Bull Math Biol* **60**, pp. 857-99.

Anderson, A. R. A. and Chaplain, M. A. J. (1997) A mathematical model for capillary network formation in the absence of endothelial cell proliferation. *App. Math. Letters* **11**, pp. 109-114.

Antonescu, C. R., Elahi, A., Healey, J. H., Brennan, M. F., Lui, M. Y., Lewis, J., Jhanwar, S. C., Woodruff, J. M. and Ladanyi, M. (2000) Monoclonality of multifocal myxoid liposarcoma: confirmation by analysis of TLS-CHOP or EWS-CHOP rearrangements. *Clin Cancer Res* **6**, pp. 2788-93.

Araujo, R. P. and McElwain, D. L. (2004) New insights into vascular collapse and growth dynamics in solid tumors. *J Theor Biol* **228**, pp. 335-46.

Armitage, P. and Doll, R. (1954) The age distribution of cancer and a multi-stage theory of carcinogenesis. *Br J Cancer* **8**, pp. 1-12.

Asahara, T., Murohara, T., Sullivan, A., Silver, M., van der Zee, R., Li, T., Witzenbichler, B., Schatteman, G. and Isner, J. M. (1997) Isolation of putative progenitor endothelial cells for angiogenesis. *Science* **275**, pp. 964-7.

Asahara, T., Takahashi, T., Masuda, H., Kalka, C., Chen, D., Iwaguro, H., Inai, Y., Silver, M. and Isner, J. M. (1999) VEGF contributes to postnatal neovascularization by mobilizing bone marrow-derived endothelial progenitor cells. *Embo J* **18**, pp. 3964-72.

Bach, S. P., Renehan, A. G. and Potten, C. S. (2000) Stem cells: the intestinal stem cell as a paradigm. *Carcinogenesis* **21**, pp. 469-76.

Bardelli, A., Cahill, D. P., Lederer, G., Speicher, M. R., Kinzler, K. W., Vogelstein, B. and Lengauer, C. (2001) Carcinogen-specific induction of genetic instability. *Proc Natl Acad Sci U S A* **98**, pp. 5770-5.

Baumann, S., Krueger, A., Kirchhoff, S. and Krammer, P. H. (2002) Regulation of T cell apoptosis during the immune response. *Curr Mol Med* **2**, pp. 257-72.

Bayko, L., Rak, J., Man, S., Bicknell, R., Ferrara, N. and Kerbel, R. S. (1998) The dormant in vivo phenotype of early stage primary human melanoma: termination by overexpression of vascular endothelial growth factor. *Angiogenesis* **2**, pp. 203-17.

Beerepoot, L. V., Mehra, N., Vermaat, J. S., Zonnenberg, B. A., Gebbink, M. F. and Voest, E. E. (2004) Increased levels of viable circulating endothelial cells are an indicator of progressive disease in cancer patients. *Ann Oncol* **15**, pp. 139-45.

Behrens, J., Jerchow, B. A., Wurtele, M., Grimm, J., Asbrand, C., Wirtz, R., Kuhl, M., Wedlich, D. and Birchmeier, W. (1998) Functional interaction of an axin homolog, conductin, with beta-catenin, APC, and GSK3beta. *Science* **280**, pp. 596-9.

Belz, G. T., Vremec, D., Febbraio, M., Corcoran, L., Shortman, K., Carbone, F. R. and Heath, W. R. (2002) CD36 is differentially expressed by CD8+ splenic dendritic cells but is not required for cross-presentation in vivo. *J Immunol* **168**, pp. 6066-70.

Blagosklonny, M. V. (2002) P53: an ubiquitous target of anticancer drugs. *Int J Cancer* **98**, pp. 161-6.

Blankenstein, T. and Schuler, T. (2002) Cross-priming versus cross-tolerance: are two signals enough? *Trends Immunol* **23**, pp. 171-3.

Bolontrade, M. F., Zhou, R. R. and Kleinerman, E. S. (2002) Vasculogenesis Plays a Role in the Growth of Ewing's Sarcoma in Vivo. *Clin Cancer Res* **8**, pp. 3622-7.

Breivik, J. and Gaudernack, G. (1999a) Carcinogenesis and natural selection: a new perspective to the genetics and epigenetics of colorectal cancer. *Adv Cancer Res* **76**, pp. 187-212.

Breivik, J. and Gaudernack, G. (1999b) Genomic instability, DNA methylation, and natural selection in colorectal carcinogenesis. *Semin Cancer Biol* **9**, pp. 245-54.

Breward, C. J., Byrne, H. M. and Lewis, C. E. (2003) A multiphase model describing vascular tumour growth. *Bull Math Biol* **65**, pp. 609-40.

Brown, D., Kogan, S., Lagasse, E., Weissman, I., Alcalay, M., Pelicci, P. G., Atwater, S. and Bishop, J. M. (1997) A PMLRARalpha transgene initiates murine acute promyelocytic leukemia. *Proc Natl Acad Sci U S A* **94**, pp. 2551-6.

Bru, A., Albertos, S., Luis Subiza, J., Garcia-Asenjo, J. L. and Bru, I. (2003) The universal dynamics of tumor growth. *Biophys J* **85**, pp. 2948-61.

Buchler, P., Reber, H. A., Ullrich, A., Shiroiki, M., Roth, M., Buchler, M. W., Lavey, R. S., Friess, H. and Hines, O. J. (2003) Pancreatic cancer growth is inhibited by blockade of VEGF-RII. *Surgery* **134**, pp. 772-82.

Budd, R. C. (2001) Activation-induced cell death. *Curr Opin Immunol* **13**, pp. 356-62.

Burnet, F. M. (1957) Cancer - a biological approach. *Brit. Med. J.* **1**.

Byrne, H. and Preziosi, L. (2003) Modelling solid tumour growth using the theory of mixtures. *Math Med Biol* **20**, pp. 341-66.

Cahill, D. P., Kinzler, K. W., Vogelstein, B. and Lengauer, C. (1999) Genetic instability and darwinian selection in tumours. *Trends Cell Biol* **9**, pp. M57-60.

Cahill, D. P., Lengauer, C., Yu, J., Riggins, G. J., Willson, J. K., Markowitz, S. D., Kinzler, K. W. and Vogelstein, B. (1998) Mutations of mitotic checkpoint genes in human cancers. *Nature* **392**, pp. 300-3.

Cairns, J. (2002) Somatic stem cells and the kinetics of mutagenesis and carcinogenesis. *Proc Natl Acad Sci U S A* **99**, pp. 10567-70.

Calabrese, P., Tavare, S. and Shibata, D. (2004) Pretumor progression: clonal evolution of human stem cell populations. *Am J Pathol* **164**, pp. 1337-46.

Campisi, J. (2001) Cellular senescence, aging and cancer. *ScientificWorldJournal* **1**, pp. 65.

Campisi, J. (2003a) Cancer and ageing: rival demons? *Nat Rev Cancer* **3**, pp. 339-49.

Campisi, J. (2003b) Cellular senescence and apoptosis: how cellular responses might influence aging phenotypes. *Exp Gerontol* **38**, pp. 5-11.

Camplejohn, R. S., Gilchrist, R., Easton, D., McKenzie-Edwards, E., Barnes, D. M., Eccles, D. M., Ardern-Jones, A., Hodgson, S. V., Duddy, P. M. and Eeles, R. A. (2003) Apoptosis, ageing and cancer susceptibility. *Br J Cancer* **88**, pp. 487-90.

Chavez-Reyes, A., Parant, J. M., Amelse, L. L., de Oca Luna, R. M., Korsmeyer, S. J. and Lozano, G. (2003) Switching mechanisms of cell death in mdm2- and mdm4-null mice by deletion of p53 downstream targets. *Cancer Res* **63**, pp. 8664-9.

Cunha, G. R. and Matrisian, L. M. (2002) It's not my fault, blame it on my microenvironment. *Differentiation* **70**, pp. 469-72.

Cunningham, J. M., Christensen, E. R., Tester, D. J., Kim, C. Y., Roche, P.

C., Burgart, L. J. and Thibodeau, S. N. (1998) Hypermethylation of the hMLH1 promoter in colon cancer with microsatellite instability. *Cancer Res* **58**, pp. 3455-60.

Daley, G. Q. (2003) Towards combination target-directed chemotherapy for chronic myeloid leukemia: role of farnesyl transferase inhibitors. *Semin Hematol* **40**, pp. 11-4.

Dallon, J. C. and Sherratt, J. A. (1998) A mathematical model for fibroblast and collagen orientation. *Bull Math Biol* **60**, pp. 101-29.

Davidoff, A. M., Ng, C. Y., Brown, P., Leary, M. A., Spurbeck, W. W., Zhou, J., Horwitz, E., Vanin, E. F. and Nienhuis, A. W. (2001) Bone marrow-derived cells contribute to tumor neovasculature and, when modified to express an angiogenesis inhibitor, can restrict tumor growth in mice. *Clin Cancer Res* **7**, pp. 2870-9.

De Boer, R. J. and Perelson, A. S. (1995) Towards a general function describing T cell proliferation. *J Theor Biol* **175**, pp. 567-76.

den Haan, J. M. and Bevan, M. J. (2001) Antigen presentation to CD8+ T cells: cross-priming in infectious diseases. *Curr Opin Immunol* **13**, pp. 437-41.

den Haan, J. M., Lehar, S. M. and Bevan, M. J. (2000) CD8(+) but not CD8(-) dendritic cells cross-prime cytotoxic T cells in vivo. *J Exp Med* **192**, pp. 1685-96.

Dix, B. R., O'Carroll, S. J., Myers, C. J., Edwards, S. J. and Braithwaite, A. W. (2000) Efficient induction of cell death by adenoviruses requires binding of E1B55k and p53. *Cancer Res* **60**, pp. 2666-72.

Donehower, L. A. (2002) Does p53 affect organismal aging? *J Cell Physiol* **192**, pp. 23-33.

Drake, C. J. (2003) Embryonic and adult vasculogenesis. *Birth Defects Res Part C Embryo Today* **69**, pp. 73-82.

Dunn, G. P., Bruce, A. T., Ikeda, H., Old, L. J. and Schreiber, R. D. (2002) Cancer immunoediting: from immunosurveillance to tumor escape. *Nat Immunol* **3**, pp. 991-8.

Eden, A., Gaudet, F., Waghmare, A. and Jaenisch, R. (2003) Chromosomal instability and tumors promoted by DNA hypomethylation. *Science* **300**, pp. 455.

Ehrlich, P. (1909) Ueber den jetzigen Stand der Karzinomforschung. *Ned. Tijdschr. Geneeskd.* **5**.

Eigen, M. and Schuster, P. (1979) *The Hypercycle. A principle of natural self organization.* Berlin: Springer.

Esteller, M., Fraga, M. F., Guo, M., Garcia-Foncillas, J., Hedenfalk, I., Godwin, A. K., Trojan, J., Vaurs-Barriere, C., Bignon, Y. J., Ramus, S., Benitez, J., Caldes, T., Akiyama, Y., Yuasa, Y., Launonen, V., Canal, M. J., Rodriguez, R., Capella, G., Peinado, M. A., Borg, A., Aaltonen, L. A., Ponder, B. A., Baylin, S. B. and Herman, J. G. (2001) DNA methylation patterns in hereditary human cancers mimic sporadic tumorigenesis. *Hum Mol Genet* **10**, pp. 3001-7.

Evans, N. D., Errington, R. J., Shelley, M., Feeney, G. P., Chapman, M. J., Godfrey, K. R., Smith, P. J. and Chappell, M. J. (2004) A mathematical model for the in vitro kinetics of the anti-cancer agent topotecan. *Math Biosci* **189**, pp. 185-217.

Evans, S. C. and Lozano, G. (1997) The Li-Fraumeni syndrome: an inherited susceptibility to cancer. *Mol Med Today* **3**, pp. 390-5.

Fauth, C. and Speicher, M. R. (2001) Classifying by colors: FISH-based genome analysis. *Cytogenet Cell Genet* **93**, pp. 1-10.

Ferreira, S. C., Jr., Martins, M. L. and Vilela, M. J. (2002) Reaction-diffusion model for the growth of avascular tumor. *Phys Rev E Stat Nonlin Soft Matter Phys* **65**, pp. 021907.

Filoche, M. and Schwartz, L. (2004) Cancer death statistics: analogy between epidemiology and critical systems in physics. *Med Hypotheses* **62**, pp. 704-9.

Finkel, T. and Holbrook, N. J. (2000) Oxidants, oxidative stress and the biology of ageing. *Nature* **408**, pp. 239-47.

Fodde, R., Kuipers, J., Rosenberg, C., Smits, R., Kielman, M., Gaspar, C., van Es, J. H., Breukel, C., Wiegant, J., Giles, R. H. and Clevers, H. (2001a) Mutations in the APC tumour suppressor gene cause chromosomal instability. *Nat Cell Biol* **3**, pp. 433-8.

Fodde, R., Smits, R. and Clevers, H. (2001b) APC, signal transduction and genetic instability in colorectal cancer. *Nat Rev Cancer* **1**, pp. 55-67.

Folkman, J. (1971) Tumor angiogenesis: therapeutic implications. *N Engl J Med* **285**, pp. 1182-6.

Folkman, J. (1995a) Angiogenesis in cancer, vascular, rheumatoid and other disease. *Nat Med* **1**, pp. 27-31.

Folkman, J. (1995b) Angiogenesis inhibitors generated by tumors. *Mol Med* **1**, pp. 120-2.

Folkman, J. (2002) Role of angiogenesis in tumor growth and metastasis. *Semin Oncol* **29**, pp. 15-8.

Folkman, J. and Kalluri, R. (2004) Cancer without disease. *Nature* **427**, pp. 787.

Frank, S. A. (2003) Somatic mosaicism and cancer: inference based on a conditional Luria-Delbruck distribution. *J Theor Biol* **223**, pp. 405-12.

Frank, S. A. (2004) Age-specific acceleration of cancer. *Curr Biol* **14**, pp. 242-6.

Frank, S. A. and Nowak, M. A. (2003) Cell biology: Developmental predisposition to cancer. *Nature* **422**, pp. 494.

Franks, S. J., Byrne, H. M., Mudhar, H. S., Underwood, J. C. and Lewis, C. E. (2003) Mathematical modelling of comedo ductal carcinoma in situ of the breast. *Math Med Biol* **20**, pp. 277-308.

Freytag, S. O., Rogulski, K. R., Paielli, D. L., Gilbert, J. D. and Kim, J. H. (1998) A novel three-pronged approach to kill cancer cells selectively: concomitant viral, double suicide gene, and radiotherapy [see comments]. *Hum Gene Ther* **9**, pp. 1323-33.

Frigyesi, A., Gisselsson, D., Hansen, G. B., Soller, M., Mitelman, F. and Hoglund, M. (2004) A model for karyotypic evolution in testicular germ cell tumors. *Genes Chromosomes Cancer* **40**, pp. 172-8.

Frigyesi, A., Gisselsson, D., Mitelman, F. and Hoglund, M. (2003) Power law distribution of chromosome aberrations in cancer. *Cancer Res* **63**, pp. 7094-

7.
Fuchs, E. and Segre, J. A. (2000) Stem cells: a new lease on life. *Cell* **100**, pp. 143-55.
Fuchs, E. J. and Matzinger, P. (1996) Is cancer dangerous to the immune system? *Semin Immunol* **8**, pp. 271-80.
Ganly, I., Kirn, D., Eckhardt, S. G., Rodriguez, G. I., Soutar, D. S., Otto, R., Robertson, A. G., Park, O., Gulley, M. L., Heise, C., Von Hoff, D. D. and Kaye, S. B. (2000) A phase I study of Onyx-015, an E1B attenuated adenovirus, administered intratumorally to patients with recurrent head and neck cancer. *Clin Cancer Res* **6**, pp. 798-806.
Gasche, C., Chang, C. L., Rhees, J., Goel, A. and Boland, C. R. (2001) Oxidative stress increases frameshift mutations in human colorectal cancer cells. *Cancer Res* **61**, pp. 7444-8.
Gatenby, R. A. (1996) Application of competition theory to tumour growth: implications for tumour biology and treatment. *Eur J Cancer* **32A**, pp. 722-6.
Gatenby, R. A. and Gawlinski, E. T. (1996) A reaction-diffusion model of cancer invasion. *Cancer Res* **56**, pp. 5745-53.
Gatenby, R. A. and Gawlinski, E. T. (2003) The glycolytic phenotype in carcinogenesis and tumor invasion: insights through mathematical models. *Cancer Res* **63**, pp. 3847-54.
Gatenby, R. A. and Vincent, T. L. (2003a) Application of quantitative models from population biology and evolutionary game theory to tumor therapeutic strategies. *Mol Cancer Ther* **2**, pp. 919-27.
Gatenby, R. A. and Vincent, T. L. (2003b) An evolutionary model of carcinogenesis. *Cancer Res* **63**, pp. 6212-20.
Gaudet, F., Hodgson, J. G., Eden, A., Jackson-Grusby, L., Dausman, J., Gray, J. W., Leonhardt, H. and Jaenisch, R. (2003) Induction of tumors in mice by genomic hypomethylation. *Science* **300**, pp. 489-92.
Gemma, A., Seike, M., Seike, Y., Uematsu, K., Hibino, S., Kurimoto, F., Yoshimura, A., Shibuya, M., Harris, C. C. and Kudoh, S. (2000) Somatic mutation of the hBUB1 mitotic checkpoint gene in primary lung cancer. *Genes Chromosomes Cancer* **29**, pp. 213-8.
Gilhar, A., Ullmann, Y., Karry, R., Shalaginov, R., Assy, B., Serafimovich, S. and Kalish, R. S. (2004) Ageing of human epidermis: the role of apoptosis, Fas and telomerase. *Br J Dermatol* **150**, pp. 56-63.
Grisolano, J. L., Wesselschmidt, R. L., Pelicci, P. G. and Ley, T. J. (1997) Altered myeloid development and acute leukemia in transgenic mice expressing PML-RAR alpha under control of cathepsin G regulatory sequences. *Blood* **89**, pp. 376-87.
Guba, M., Cernaianu, G., Koehl, G., Geissler, E. K., Jauch, K. W., Anthuber, M., Falk, W. and Steinbauer, M. (2001) A primary tumor promotes dormancy of solitary tumor cells before inhibiting angiogenesis. *Cancer Res* **61**, pp. 5575-9.
Guillemard, V. and Saragovi, H. U. (2004) Novel approaches for targeted cancer therapy. *Curr Cancer Drug Targets* **4**, pp. 313-26.
Guilloux, Y., Bai, X. F., Liu, X., Zheng, P. and Liu, Y. (2001) Optimal induction

of effector but not memory antitumor cytotoxic T lymphocytes involves direct antigen presentation by the tumor cells. *Cancer Res* **61**, pp. 1107-12.

Guiot, C., Degiorgis, P. G., Delsanto, P. P., Gabriele, P. and Deisboeck, T. S. (2003) Does tumor growth follow a "universal law"? *J Theor Biol* **225**, pp. 147-51.

Hafner, C., Knuechel, R., Stoehr, R. and Hartmann, A. (2002) Clonality of multifocal urothelial carcinomas: 10 years of molecular genetic studies. *Int J Cancer* **101**, pp. 1-6.

Hahnfeldt, P., Panigrahy, D., Folkman, J. and Hlatky, L. (1999a) Tumor development under angiogenic signaling: a dynamical theory of tumor growth, treatment response, and postvascular dormancy. *Cancer Res* **59**, pp. 4770-5.

Hahnfeldt, P., Panigrahy, D., Folkman, J. and Hlatky, L. (1999b) Tumor development under angiogenic signaling: a dynamical theory of tumor growth, treatment response, and postvascular dormancy. *Cancer Res* **59**, pp. 4770-5.

Hall, A. R., Dix, B. R., O'Carroll, S. J. and Braithwaite, A. W. (1998) p53-dependent cell death/apoptosis is required for a productive adenovirus infection [see comments]. *Nat Med* **4**, pp. 1068-72.

Harrison, D., Sauthoff, H., Heitner, S., Jagirdar, J., Rom, W. N. and Hay, J. G. (2001) Wild-type adenovirus decreases tumor xenograft growth, but despite viral persistence complete tumor responses are rarely achieved– deletion of the viral E1b-19-kD gene increases the viral oncolytic effect. *Hum Gene Ther* **12**, pp. 1323-32.

Hartmann, A., Rosner, U., Schlake, G., Dietmaier, W., Zaak, D., Hofstaedter, F. and Knuechel, R. (2000) Clonality and genetic divergence in multifocal low-grade superficial urothelial carcinoma as determined by chromosome 9 and p53 deletion analysis. *Lab Invest* **80**, pp. 709-18.

Hasan, S., Hassan, M., Oke, L., Dinh, K., Onojobi, G., Lombardo, F., Dawkins, F. and Jack, M. (2003) Advanced gastrointestinal stromal tumors successfully treated with imatinib mesylate: a report of two cases. *J Natl Med Assoc* **95**, pp. 1208-10.

Hasty, P., Campisi, J., Hoeijmakers, J., van Steeg, H. and Vijg, J. (2003) Aging and genome maintenance: lessons from the mouse? *Science* **299**, pp. 1355-9.

Haylock, R. G. and Muirhead, C. R. (2004) Fitting the two-stage model of carcinogenesis to nested case-control data on the Colorado Plateau uranium miners: dependence on data assumptions. *Radiat Environ Biophys* **42**, pp. 257-63.

He, L. Z., Tribioli, C., Rivi, R., Peruzzi, D., Pelicci, P. G., Soares, V., Cattoretti, G. and Pandolfi, P. P. (1997) Acute leukemia with promyelocytic features in PML/RARalpha transgenic mice. *Proc Natl Acad Sci U S A* **94**, pp. 5302-7.

Heath, W. R. and Carbone, F. R. (2001a) Cross-presentation in viral immunity and self-tolerance. *Nat Rev Immunol* **1**, pp. 126-34.

Heath, W. R. and Carbone, F. R. (2001b) Cross-presentation, dendritic cells, tolerance and immunity. *Annu Rev Immunol* **19**, pp. 47-64.

Heise, C., Sampson-Johannes, A., Williams, A., McCormick, F., Von Hoff, D. D. and Kirn, D. H. (1997) ONYX-015, an E1B gene-attenuated adenovirus, causes tumor-specific cytolysis and antitumoral efficacy that can be augmented by standard chemotherapeutic agents [see comments]. *Nat Med* **3**, pp. 639-45.

Heise, C. C., Williams, A., Olesch, J. and Kirn, D. H. (1999a) Efficacy of a replication-competent adenovirus (ONYX-015) following intratumoral injection: intratumoral spread and distribution effects. *Cancer Gene Ther* **6**, pp. 499-504.

Heise, C. C., Williams, A. M., Xue, S., Propst, M. and Kirn, D. H. (1999b) Intravenous administration of ONYX-015, a selectively replicating adenovirus, induces antitumoral efficacy. *Cancer Res* **59**, pp. 2623-8.

Hildeman, D. A., Zhu, Y., Mitchell, T. C., Kappler, J. and Marrack, P. (2002) Molecular mechanisms of activated T cell death in vivo. *Curr Opin Immunol* **14**, pp. 354-9.

Hoglund, M., Gisselsson, D., Hansen, G. B., Sall, T. and Mitelman, F. (2002a) Multivariate analysis of chromosomal imbalances in breast cancer delineates cytogenetic pathways and reveals complex relationships among imbalances. *Cancer Res* **62**, pp. 2675-80.

Hoglund, M., Gisselsson, D., Hansen, G. B., Sall, T., Mitelman, F. and Nilbert, M. (2002b) Dissecting karyotypic patterns in colorectal tumors: two distinct but overlapping pathways in the adenoma-carcinoma transition. *Cancer Res* **62**, pp. 5939-46.

Hoglund, M., Sall, T., Heim, S., Mitelman, F., Mandahl, N. and Fadl-Elmula, I. (2001) Identification of cytogenetic subgroups and karyotypic pathways in transitional cell carcinoma. *Cancer Res* **61**, pp. 8241-6.

Holland, E. C. (2000) Glioblastoma multiforme: the terminator. *Proc Natl Acad Sci U S A* **97**, pp. 6242-4.

Hsu, M. Y., Meier, F. and Herlyn, M. (2002) Melanoma development and progression: a conspiracy between tumor and host. *Differentiation* **70**, pp. 522-36.

Huang, J., Papadopoulos, N., McKinley, A. J., Farrington, S. M., Curtis, L. J., Wyllie, A. H., Zheng, S., Willson, J. K., Markowitz, S. D., Morin, P., Kinzler, K. W., Vogelstein, B. and Dunlop, M. G. (1996) APC mutations in colorectal tumors with mismatch repair deficiency. *Proc Natl Acad Sci U S A* **93**, pp. 9049-54.

Hurlstone, D. P., Cross, S. S., Adam, I., Shorthouse, A. J., Brown, S., Sanders, D. S. and Lobo, A. J. (2004) Efficacy of high magnification chromoscopic colonoscopy for the diagnosis of neoplasia in flat and depressed lesions of the colorectum: a prospective analysis. *Gut* **53**, pp. 284-90.

Imai, Y., Shiratori, Y., Kato, N., Inoue, T. and Omata, M. (1999) Mutational inactivation of mitotic checkpoint genes, hsMAD2 and hBUB1, is rare in sporadic digestive tract cancers. *Jpn J Cancer Res* **90**, pp. 837-40.

Itahana, K., Dimri, G. and Campisi, J. (2001) Regulation of cellular senescence by p53. *Eur J Biochem* **268**, pp. 2784-91.

Iwasa, Y., Michor, F. and Nowak, M. A. (2004) Stochastic tunnels in evolutionary dynamics. *Genetics* **166**, pp. 1571-9.

Jackson, A. L. and Loeb, L. A. (1998) The mutation rate and cancer. *Genetics* **148**, pp. 1483-90.

Jackson, A. L. and Loeb, L. A. (2001) The contribution of endogenous sources of DNA damage to the multiple mutations in cancer. *Mutat Res* **477**, pp. 7-21.

Janeway, C., P., T., M., W. and D., C. J. (1999) *Immunobiology: The immune system in health and disease.* New York: Current Biology Ltd.

Janeway, C. A., Jr. (2002) A trip through my life with an immunological theme. *Annu Rev Immunol* **20**, pp. 1-28.

Jansen-Durr, P. (2002) Cell death and ageing: a question of cell type. *Scientific-WorldJournal* **2**, pp. 943-8.

John, A. M., Thomas, N. S., Mufti, G. J. and Padua, R. A. (2004) Targeted therapies in myeloid leukemia. *Semin Cancer Biol* **14**, pp. 41-62.

Joseph-Silverstein, J. and Silverstein, R. L. (1998) Cell adhesion molecules: an overview. *Cancer Invest* **16**, pp. 176-82.

Junker, K., Schlichter, A., Hindermann, W. and Schubert, J. (1999) Genetic characterization of multifocal tumor growth in renal cell carcinoma. *Kidney Int* **56**, pp. 1291-4.

Junker, K., Schlichter, A., Junker, U., Knofel, B., Kosmehl, H., Schubert, J. and Claussen, U. (1997) Cytogenetic, histopathologic, and immunologic studies of multifocal renal cell carcinoma. *Cancer* **79**, pp. 975-81.

Junker, K., Thrum, K., Schlichter, A., Muller, G., Hindermann, W. and Schubert, J. (2002) Clonal origin of multifocal renal cell carcinoma as determined by microsatellite analysis. *J Urol* **168**, pp. 2632-6.

Kahlem, P., Dorken, B. and Schmitt, C. A. (2004) Cellular senescence in cancer treatment: friend or foe? *J Clin Invest* **113**, pp. 169-74.

Kane, M. F., Loda, M., Gaida, G. M., Lipman, J., Mishra, R., Goldman, H., Jessup, J. M. and Kolodner, R. (1997) Methylation of the hMLH1 promoter correlates with lack of expression of hMLH1 in sporadic colon tumors and mismatch repair-defective human tumor cell lines. *Cancer Res* **57**, pp. 808-11.

Kansal, A. R., Torquato, S., Harsh, G. I., Chiocca, E. A. and Deisboeck, T. S. (2000) Simulated brain tumor growth dynamics using a three-dimensional cellular automaton. *J Theor Biol* **203**, pp. 367-82.

Kaplan, K. B., Burds, A. A., Swedlow, J. R., Bekir, S. S., Sorger, P. K. and Nathke, I. S. (2001) A role for the Adenomatous Polyposis Coli protein in chromosome segregation. *Nat Cell Biol* **3**, pp. 429-32.

Katoh, M. (2003) WNT2 and human gastrointestinal cancer (review). *Int J Mol Med* **12**, pp. 811-6.

Keller, J. J., Offerhaus, G. J., Drillenburg, P., Caspers, E., Musler, A., Ristimaki, A. and Giardiello, F. M. (2001) Molecular analysis of sulindac-resistant adenomas in familial adenomatous polyposis. *Clin Cancer Res* **7**, pp. 4000-7.

Kemp, C. J., Donehower, L. A., Bradley, A. and Balmain, A. (1993) Reduction of p53 gene dosage does not increase initiation or promotion but enhances malignant progression of chemically induced skin tumors. *Cell* **74**, pp. 813-22.

Khuri, F. R., Nemunaitis, J., Ganly, I., Arseneau, J., Tannock, I. F., Romel, L., Gore, M., Ironside, J., MacDougall, R. H., Heise, C., Randlev, B., Gillenwater, A. M., Bruso, P., Kaye, S. B., Hong, W. K. and Kirn, D. H. (2000) a controlled trial of intratumoral ONYX-015, a selectively-replicating adenovirus, in combination with cisplatin and 5-fluorouracil in patients with recurrent head and neck cancer [see comments]. *Nat Med* **6**, pp. 879-85.

Kim, H. K., Song, K. S., Kim, H. O., Chung, J. H., Lee, K. R., Lee, Y. J., Lee, D. H., Lee, E. S., Ryu, K. W. and Bae, J. M. (2003) Circulating numbers of endothelial progenitor cells in patients with gastric and breast cancer. *Cancer Lett* **198**, pp. 83-8.

Kim, K. J., Li, B., Winer, J., Armanini, M., Gillett, N., Phillips, H. S. and Ferrara, N. (1993) Inhibition of vascular endothelial growth factor-induced angiogenesis suppresses tumour growth in vivo. *Nature* **362**, pp. 841-4.

Kim, K. M., Calabrese, P., Tavare, S. and Shibata, D. (2004) Enhanced stem cell survival in familial adenomatous polyposis. *Am J Pathol* **164**, pp. 1369-77.

Kim, K. M. and Shibata, D. (2002) Methylation reveals a niche: stem cell succession in human colon crypts. *Oncogene* **21**, pp. 5441-9.

Kimura, M. (1994) *Population Genetics, Molecular Evolution, and Neutral Theory: Selected Papers.* Chicago: University of Chicago Press.

Kinzler, K. W. and Vogelstein, B. (1998) *The Genetic Basis of Cancer.* Toronto: McGraw-HIll.

Kirchner, T., Muller, S., Hattori, T., Mukaisyo, K., Papadopoulos, T., Brabletz, T. and Jung, A. (2001) Metaplasia, intraepithelial neoplasia and early cancer of the stomach are related to dedifferentiated epithelial cells defined by cytokeratin-7 expression in gastritis. *Virchows Arch* **439**, pp. 512-22.

Kirkwood, T. B. (2002) p53 and ageing: too much of a good thing? *Bioessays* **24**, pp. 577-9.

Kirn, D., Hermiston, T. and McCormick, F. (1998) ONYX-015: clinical data are encouraging [letter; comment]. *Nat Med* **4**, pp. 1341-2.

Kirn, D. H. and McCormick, F. (1996) Replicating viruses as selective cancer therapeutics. *Mol Med Today* **2**, pp. 519-27.

Kirschner, D. and Panetta, J. C. (1998) Modeling immunotherapy of the tumor-immune interaction. *J Math Biol* **37**, pp. 235-52.

Knudson, A. G. (1996) Hereditary cancer: two hits revisited. *J Cancer Res Clin Oncol* **122**, pp. 135-40.

Knudson, A. G., Jr. (1971) Mutation and cancer: statistical study of retinoblastoma. *Proc Natl Acad Sci U S A* **68**, pp. 820-3.

Kolodner, R. D., Putnam, C. D. and Myung, K. (2002) Maintenance of genome stability in Saccharomyces cerevisiae. *Science* **297**, pp. 552-7.

Komarova, N. (2004) Does Cancer Solve an Optimization Problem? *Cell Cycle* **3**.

Komarova, N. L., Lengauer, C., Vogelstein, B. and Nowak, M. A. (2002) Dynamics of genetic instability in sporadic and familial colorectal cancer. *Cancer Biol Ther* **1**, pp. 685-92.

Komarova, N. L. and Mironov, V. (2004) Angiogenesis vs vasculogenesis in tumor growth: identifying the signature of the dominant process. *submitted*

Komarova, N. L., Sengupta, A. and Nowak, M. A. (2003) Mutation-selection networks of cancer initiation: tumor suppressor genes and chromosomal instability. *J Theor Biol* **223**, pp. 433-50.

Komarova, N. L. and Wang, L. (2004) Initiation of colorectal cancer: where do the two hits hit? *Cell Cycle* **in press**.

Komarova, N. L. and Wodarz, D. (2003) Evolutionary dynamics of mutator phenotypes in cancer: implications for chemotherapy. *Cancer Res* **63**, pp. 6635-42.

Komarova, N. L. and Wodarz, D. (2004) The optimal rate of chromosome loss for the inactivation of tumor suppressor genes in cancer. *Proc Natl Acad Sci U S A* **101**, pp. 7017-21.

Krewski, D., Zielinski, J. M., Hazelton, W. D., Garner, M. J. and Moolgavkar, S. H. (2003) The use of biologically based cancer risk models in radiation epidemiology. *Radiat Prot Dosimetry* **104**, pp. 367-76.

Kupryjanczyk, J., Thor, A. D., Beauchamp, R., Poremba, C., Scully, R. E. and Yandell, D. W. (1996) Ovarian, peritoneal, and endometrial serous carcinoma: clonal origin of multifocal disease. *Mod Pathol* **9**, pp. 166-73.

Kuukasjarvi, T., Karhu, R., Tanner, M., Kahkonen, M., Schaffer, A., Nupponen, N., Pennanen, S., Kallioniemi, A., Kallioniemi, O. P. and Isola, J. (1997a) Genetic heterogeneity and clonal evolution underlying development of asynchronous metastasis in human breast cancer. *Cancer Res* **57**, pp. 1597-604.

Kuukasjarvi, T., Tanner, M., Pennanen, S., Karhu, R., Kallioniemi, O. P. and Isola, J. (1997b) Genetic changes in intraductal breast cancer detected by comparative genomic hybridization. *Am J Pathol* **150**, pp. 1465-71.

Laird, A. K. (1969) Dynamics of growth in tumors and in normal organisms. *Natl Cancer Inst Monogr* **30**, pp. 15-28.

Lamlum, H., Ilyas, M., Rowan, A., Clark, S., Johnson, V., Bell, J., Frayling, I., Efstathiou, J., Pack, K., Payne, S., Roylance, R., Gorman, P., Sheer, D., Neale, K., Phillips, R., Talbot, I., Bodmer, W. and Tomlinson, I. (1999) The type of somatic mutation at APC in familial adenomatous polyposis is determined by the site of the germline mutation: a new facet to Knudson's 'two-hit' hypothesis. *Nat Med* **5**, pp. 1071-5.

Lamprecht, S. A. and Lipkin, M. (2002) Migrating colonic crypt epithelial cells: primary targets for transformation. *Carcinogenesis* **23**, pp. 1777-80.

Lazareff, J. A., Suwinski, R., De Rosa, R. and Olmstead, C. E. (1999) Tumor volume and growth kinetics in hypothalamic-chiasmatic pediatric low grade gliomas. *Pediatr Neurosurg* **30**, pp. 312-9.

Lengauer, C., Kinzler, K. W. and Vogelstein, B. (1997) Genetic instability in colorectal cancers. *Nature* **386**, pp. 623-7.

Lengauer, C., Kinzler, K. W. and Vogelstein, B. (1998) Genetic instabilities in human cancers. *Nature* **396**, pp. 643-9.

Li, Y. and Benezra, R. (1996) Identification of a human mitotic checkpoint gene: hsMAD2. *Science* **274**, pp. 246-8.

Lindblom, A. (2001) Different mechanisms in the tumorigenesis of proximal and distal colon cancers. *Curr Opin Oncol* **13**, pp. 63-9.

Loeb, K. R. and Loeb, L. A. (2000) Significance of multiple mutations in cancer. *Carcinogenesis* **21**, pp. 379-85.

Loeb, L. A. (1991) Mutator phenotype may be required for multistage carcino-

genesis. *Cancer Res* **51**, pp. 3075-9.
Loeb, L. A. (2001) A mutator phenotype in cancer. *Cancer Res* **61**, pp. 3230-9.
Loeb, L. A., Springgate, C. F. and Battula, N. (1974) Errors in DNA replication as a basis of malignant changes. *Cancer Res* **34**, pp. 2311-21.
Louhelainen, J., Wijkstrom, H. and Hemminki, K. (2000) Allelic losses demonstrate monoclonality of multifocal bladder tumors. *Int J Cancer* **87**, pp. 522-7.
Luebeck, E. G. and Moolgavkar, S. H. (2002) Multistage carcinogenesis and the incidence of colorectal cancer. *Proc Natl Acad Sci U S A* **99**, pp. 15095-100.
Luo, G., Santoro, I. M., McDaniel, L. D., Nishijima, I., Mills, M., Youssoufian, H., Vogel, H., Schultz, R. A. and Bradley, A. (2000) Cancer predisposition caused by elevated mitotic recombination in Bloom mice. *Nat Genet* **26**, pp. 424-9.
Lyden, D., Hattori, K., Dias, S., Costa, C., Blaikie, P., Butros, L., Chadburn, A., Heissig, B., Marks, W., Witte, L., Wu, Y., Hicklin, D., Zhu, Z., Hackett, N. R., Crystal, R. G., Moore, M. A., Hajjar, K. A., Manova, K., Benezra, R. and Rafii, S. (2001) Impaired recruitment of bone-marrow-derived endothelial and hematopoietic precursor cells blocks tumor angiogenesis and growth. *Nat Med* **7**, pp. 1194-201.
Macleod, K. (2000) Tumor suppressor genes. *Curr Opin Genet Dev* **10**, pp. 81-93.
Mandonnet, E., Delattre, J. Y., Tanguy, M. L., Swanson, K. R., Carpentier, A. F., Duffau, H., Cornu, P., Van Effenterre, R., Alvord, E. C., Jr. and Capelle, L. (2003) Continuous growth of mean tumor diameter in a subset of grade II gliomas. *Ann Neurol* **53**, pp. 524-8.
Mansury, Y. and Deisboeck, T. S. (2003) The impact of "search precision" in an agent-based tumor model. *J Theor Biol* **224**, pp. 325-37.
Marsh, D. and Zori, R. (2002) Genetic insights into familial cancers– update and recent discoveries. *Cancer Lett* **181**, pp. 125-64.
Matzavinos, A., Chaplain, M. A. and Kuznetsov, V. A. (2004) Mathematical modelling of the spatio-temporal response of cytotoxic T-lymphocytes to a solid tumour. *Math Med Biol* **21**, pp. 1-34.
Matzinger, P. (1998) An innate sense of danger. *Semin Immunol* **10**, pp. 399-415.
McCormick, F. (2003) Cancer-specific viruses and the development of ONYX-015. *Cancer Biol Ther* **2**, pp. S157-60.
Menigatti, M., Di Gregorio, C., Borghi, F., Sala, E., Scarselli, A., Pedroni, M., Foroni, M., Benatti, P., Roncucci, L., Ponz de Leon, M. and Percesepe, A. (2001) Methylation pattern of different regions of the MLH1 promoter and silencing of gene expression in hereditary and sporadic colorectal cancer. *Genes Chromosomes Cancer* **31**, pp. 357-61.
Michel, L. S., Liberal, V., Chatterjee, A., Kirchwegger, R., Pasche, B., Gerald, W., Dobles, M., Sorger, P. K., Murty, V. V. and Benezra, R. (2001) MAD2 haplo-insufficiency causes premature anaphase and chromosome instability in mammalian cells. *Nature* **409**, pp. 355-9.
Michor, F., Frank, S. A., May, R. M., Iwasa, Y. and Nowak, M. A. (2003a) Somatic selection for and against cancer. *J Theor Biol* **225**, pp. 377-82.
Michor, F., Nowak, M. A., Frank, S. A. and Iwasa, Y. (2003b) Stochastic elimi-

nation of cancer cells. *Proc R Soc Lond B Biol Sci* **270**, pp. 2017-24.

Middleton, L. P., Vlastos, G., Mirza, N. Q., Eva, S. and Sahin, A. A. (2002) Multicentric mammary carcinoma: evidence of monoclonal proliferation. *Cancer* **94**, pp. 1910-6.

Mitelman, F. (2000) Recurrent chromosome aberrations in cancer. *Mutat Res* **462**, pp. 247-53.

Miyake, H., Nakamura, H., Hara, I., Gohji, K., Arakawa, S., Kamidono, S. and Saya, H. (1998) Multifocal renal cell carcinoma: evidence for a common clonal origin. *Clin Cancer Res* **4**, pp. 2491-4.

Moolgavkar, S. H. (1978) The multistage theory of carcinogenesis and the age distribution of cancer in man. *J Natl Cancer Inst* **61**, pp. 49-52.

Moolgavkar, S. H., Day, N. E. and Stevens, R. G. (1980) Two-stage model for carcinogenesis: Epidemiology of breast cancer in females. *J Natl Cancer Inst* **65**, pp. 559-69.

Moolgavkar, S. H., Dewanji, A. and Venzon, D. J. (1988) A stochastic two-stage model for cancer risk assessment. I. The hazard function and the probability of tumor. *Risk Anal* **8**, pp. 383-92.

Moolgavkar, S. H. and Knudson, A. G., Jr. (1981) Mutation and cancer: a model for human carcinogenesis. *J Natl Cancer Inst* **66**, pp. 1037-52.

Moore, H. and Li, N. K. (2004) A mathematical model for chronic myelogenous leukemia (CML) and T cell interaction. *J Theor Biol* **227**, pp. 513-23.

Mueller, M. M. and Fusenig, N. E. (2002) Tumor-stroma interactions directing phenotype and progression of epithelial skin tumor cells. *Differentiation* **70**, pp. 486-97.

Noguchi, S., Aihara, T., Koyama, H., Motomura, K., Inaji, H. and Imaoka, S. (1994) Discrimination between multicentric and multifocal carcinomas of the breast through clonal analysis. *Cancer* **74**, pp. 872-7.

Norton, L. (1988) A Gompertzian model of human breast cancer growth. *Cancer Res* **48**, pp. 7067-71.

Nowak, M. A., Michor, F. and Iwasa, Y. (2003) The linear process of somatic evolution. *Proc Natl Acad Sci U S A* **100**, pp. 14966-9.

Offer, H., Erez, N., Zurer, I., Tang, X., Milyavsky, M., Goldfinger, N. and Rotter, V. (2002) The onset of p53-dependent DNA repair or apoptosis is determined by the level of accumulated damaged DNA. *Carcinogenesis* **23**, pp. 1025-32.

Ohshima, K., Haraoka, S., Yoshioka, S., Hamasaki, M., Fujiki, T., Suzumiya, J., Kawasaki, C., Kanda, M. and Kikuchi, M. (2000) Mutation analysis of mitotic checkpoint genes (hBUB1 and hBUBR1) and microsatellite instability in adult T-cell leukemia/lymphoma. *Cancer Lett* **158**, pp. 141-50.

Oliff, A., Gibbs, J. B. and McCormick, F. (1996) New molecular targets for cancer therapy. *Sci Am* **275**, pp. 144-9.

O'Reilly, M. S., Boehm, T., Shing, Y., Fukai, N., Vasios, G., Lane, W. S., Flynn, E., Birkhead, J. R., Olsen, B. R. and Folkman, J. (1997) Endostatin: an endogenous inhibitor of angiogenesis and tumor growth. *Cell* **88**, pp. 277-85.

O'Reilly, M. S., Holmgren, L., Chen, C. and Folkman, J. (1996) Angiostatin induces and sustains dormancy of human primary tumors in mice. *Nat Med*

2, pp. 689-92.
Oren, M. (2003) Decision making by p53: life, death and cancer. *Cell Death Differ* **10**, pp. 431-42.
Owen, M. R., Byrne, H. M. and Lewis, C. E. (2004) .Mathematical modelling of the use of macrophages as vehicles for drug delivery to hypoxic tumour sites. *J Theor Biol* **226**, pp. 377-91.
Parrinello, S., Samper, E., Krtolica, A., Goldstein, J., Melov, S. and Campisi, J. (2003) Oxygen sensitivity severely limits the replicative lifespan of murine fibroblasts. *Nat Cell Biol* **5**, pp. 741-7.
Parzen, E. (1962) *Stochastic processes.* San Francisco: Holden-Day.
Perez-Losada, J. and Balmain, A. (2003) Stem-cell hierarchy in skin cancer. *Nat Rev Cancer* **3**, pp. 434-43.
Plate, K. H., Breier, G., Weich, H. A. and Risau, W. (1992) Vascular endothelial growth factor is a potential tumour angiogenesis factor in human gliomas in vivo. *Nature* **359**, pp. 845-8.
Polakis, P. (1997) The adenomatous polyposis coli (APC) tumor suppressor. *Biochim Biophys Acta* **1332**, pp. F127-47.
Polakis, P. (1999) The oncogenic activation of beta-catenin. *Curr Opin Genet Dev* **9**, pp. 15-21.
Potten, C. S., Booth, C. and Hargreaves, D. (2003) The small intestine as a model for evaluating adult tissue stem cell drug targets. *Cell Prolif* **36**, pp. 115-29.
Potten, C. S., Kellett, M., Roberts, S. A., Rew, D. A. and Wilson, G. D. (1992) Measurement of in vivo proliferation in human colorectal mucosa using bromodeoxyuridine. *Gut* **33**, pp. 71-8.
Potten, C. S. and Loeffler, M. (1990) Stem cells: attributes, cycles, spirals, pitfalls and uncertainties. Lessons for and from the crypt. *Development* **110**, pp. 1001-20.
Preston, S. L., Wong, W. M., Chan, A. O., Poulsom, R., Jeffery, R., Goodlad, R. A., Mandir, N., Elia, G., Novelli, M., Bodmer, W. F., Tomlinson, I. P. and Wright, N. A. (2003) Bottom-up histogenesis of colorectal adenomas: origin in the monocryptal adenoma and initial expansion by crypt fission. *Cancer Res* **63**, pp. 3819-25.
Preziosi, L. (ed.) 2003 Cancer Modeling and Simulation. Mathematical Biology & Medicine Series: Chapman & Hall.
Rabbany, S. Y., Heissig, B., Hattori, K. and Rafii, S. (2003) Molecular pathways regulating mobilization of marrow-derived stem cells for tissue revascularization. *Trends Mol Med* **9**, pp. 109-17.
Rafii, S., Avecilla, S., Shmelkov, S., Shido, K., Tejada, R., Moore, M. A., Heissig, B. and Hattori, K. (2003) Angiogenic factors reconstitute hematopoiesis by recruiting stem cells from bone marrow microenvironment. *Ann N Y Acad Sci* **996**, pp. 49-60.
Rajagopalan, H., Jallepalli, P. V., Rago, C., Velculescu, V. E., Kinzler, K. W., Vogelstein, B. and Lengauer, C. (2004) Inactivation of hCDC4 can cause chromosomal instability. *Nature* **428**, pp. 77-81.
Ramanujan, S., Koenig, G. C., Padera, T. P., Stoll, B. R. and Jain, R. K. (2000) Local imbalance of proangiogenic and antiangiogenic factors: a po-

tential mechanism of focal necrosis and dormancy in tumors. *Cancer Res* **60**, pp. 1442-8.

Retsky, M. W., Swartzendruber, D. E., Wardwell, R. H. and Bame, P. D. (1990) Is Gompertzian or exponential kinetics a valid description of individual human cancer growth? *Med Hypotheses* **33**, pp. 95-106.

Ribatti, D., Vacca, A. and Dammacco, F. (2003) New non-angiogenesis dependent pathways for tumour growth. *Eur J Cancer* **39**, pp. 1835-41.

Risau, W. (1997) Mechanisms of angiogenesis. *Nature* **386**, pp. 671-4.

Risau, W., Sariola, H., Zerwes, H. G., Sasse, J., Ekblom, P., Kemler, R. and Doetschman, T. (1988) Vasculogenesis and angiogenesis in embryonic-stem-cell-derived embryoid bodies. *Development* **102**, pp. 471-8.

Ro, S. and Rannala, B. (2001) Methylation patterns and mathematical models reveal dynamics of stem cell turnover in the human colon. *Proc Natl Acad Sci U S A* **98**, pp. 10519-21.

Rogulski, K. R., Freytag, S. O., Zhang, K., Gilbert, J. D., Paielli, D. L., Kim, J. H., Heise, C. C. and Kirn, D. H. (2000) In vivo antitumor activity of ONYX-015 is influenced by p53 status and is augmented by radiotherapy. *Cancer Res* **60**, pp. 1193-6.

Rosenthal, A. N., Ryan, A., Hopster, D. and Jacobs, I. J. (2002) Molecular evidence of a common clonal origin and subsequent divergent clonal evolution in vulval intraepithelial neoplasia, vulval squamous cell carcinoma and lymph node metastases. *Int J Cancer* **99**, pp. 549-54.

Rowan, A. J., Lamlum, H., Ilyas, M., Wheeler, J., Straub, J., Papadopoulou, A., Bicknell, D., Bodmer, W. F. and Tomlinson, I. P. (2000) APC mutations in sporadic colorectal tumors: A mutational "hotspot" and interdependence of the "two hits". *Proc Natl Acad Sci U S A* **97**, pp. 3352-7.

Rubio, C. A., Kumagai, J., Kanamori, T., Yanagisawa, A., Nakamura, K. and Kato, Y. (1995) Flat adenomas and flat adenocarcinomas of the colorectal mucosa in Japanese and Swedish patients. Comparative histologic study. *Dis Colon Rectum* **38**, pp. 1075-9.

Ruijter, E. T., Miller, G. J., van de Kaa, C. A., van Bokhoven, A., Bussemakers, M. J., Debruyne, F. M., Ruiter, D. J. and Schalken, J. A. (1999) Molecular analysis of multifocal prostate cancer lesions. *J Pathol* **188**, pp. 271-7.

Schmitt, C. A. (2003) Senescence, apoptosis and therapy–cutting the lifelines of cancer. *Nat Rev Cancer* **3**, pp. 286-95.

Schuch, G., Heymach, J. V., Nomi, M., Machluf, M., Force, J., Atala, A., Eder, J. P., Jr., Folkman, J. and Soker, S. (2003) Endostatin inhibits the vascular endothelial growth factor-induced mobilization of endothelial progenitor cells. *Cancer Res* **63**, pp. 8345-50.

Schuch, G., Kisker, O., Atala, A. and Soker, S. (2002) Pancreatic tumor growth is regulated by the balance between positive and negative modulators of angiogenesis. *Angiogenesis* **5**, pp. 181-90.

Sen, S. (2000) Aneuploidy and cancer. *Curr Opin Oncol* **12**, pp. 82-8.

Seoane, J., Le, H. V. and Massague, J. (2002) Myc suppression of the p21(Cip1) Cdk inhibitor influences the outcome of the p53 response to DNA damage. *Nature* **419**, pp. 729-34.

Sharpless, N. E. and DePinho, R. A. (2002) p53: good cop/bad cop. *Cell* **110**, pp. 9-12.
Sherratt, J. A. and Chaplain, M. A. (2001) A new mathematical model for avascular tumour growth. *J Math Biol* **43**, pp. 291-312.
Shih, I. M., Wang, T. L., Traverso, G., Romans, K., Hamilton, S. R., Ben-Sasson, S., Kinzler, K. W. and Vogelstein, B. (2001a) Top-down morphogenesis of colorectal tumors. *Proc Natl Acad Sci U S A* **98**, pp. 2640-5.
Shih, I. M., Zhou, W., Goodman, S. N., Lengauer, C., Kinzler, K. W. and Vogelstein, B. (2001b) Evidence that genetic instability occurs at an early stage of colorectal tumorigenesis. *Cancer Res* **61**, pp. 818-22.
Shirakawa, K., Furuhata, S., Watanabe, I., Hayase, H., Shimizu, A., Ikarashi, Y., Yoshida, T., Terada, M., Hashimoto, D. and Wakasugi, H. (2002) Induction of vasculogenesis in breast cancer models. *Br J Cancer* **87**, pp. 1454-61.
Shweiki, D., Itin, A., Soffer, D. and Keshet, E. (1992) Vascular endothelial growth factor induced by hypoxia may mediate hypoxia-initiated angiogenesis. *Nature* **359**, pp. 843-5.
Simon, R., Eltze, E., Schafer, K. L., Burger, H., Semjonow, A., Hertle, L., Dockhorn-Dworniczak, B., Terpe, H. J. and Bocker, W. (2001) Cytogenetic analysis of multifocal bladder cancer supports a monoclonal origin and intraepithelial spread of tumor cells. *Cancer Res* **61**, pp. 355-62.
Smith, J. K., Mamoon, N. M. and Duhe, R. J. (2004) Emerging roles of targeted small molecule protein-tyrosine kinase inhibitors in cancer therapy. *Oncol Res* **14**, pp. 175-225.
Sole, R. V. and Deisboeck, T. S. (2004) An error catastrophe in cancer? *J Theor Biol* **228**, pp. 47-54.
Stoll, B. R., Migliorini, C., Kadambi, A., Munn, L. L. and Jain, R. K. (2003) A mathematical model of the contribution of endothelial progenitor cells to angiogenesis in tumors: implications for antiangiogenic therapy. *Blood* **102**, pp. 2555-61.
Takahashi, T., Kalka, C., Masuda, H., Chen, D., Silver, M., Kearney, M., Magner, M., Isner, J. M. and Asahara, T. (1999) Ischemia- and cytokine-induced mobilization of bone marrow-derived endothelial progenitor cells for neovascularization. *Nat Med* **5**, pp. 434-8.
Teixeira, M. R., Pandis, N. and Heim, S. (2003) Multicentric mammary carcinoma: evidence of monoclonal proliferation. *Cancer* **97**, pp. 715-7; author reply 717.
Thiagalingam, S., Laken, S., Willson, J. K., Markowitz, S. D., Kinzler, K. W., Vogelstein, B. and Lengauer, C. (2001) Mechanisms underlying losses of heterozygosity in human colorectal cancers. *Proc Natl Acad Sci U S A* **98**, pp. 2698-702.
Tischfield, J. A. and Shao, C. (2003) Somatic recombination redux. *Nat Genet* **33**, pp. 5-6.
Tlsty, T. D. (2001) Stromal cells can contribute oncogenic signals. *Semin Cancer Biol* **11**, pp. 97-104.
Tlsty, T. D. and Hein, P. W. (2001) Know thy neighbor: stromal cells can contribute oncogenic signals. *Curr Opin Genet Dev* **11**, pp. 54-9.

Tomlinson, I. (2000) Different pathways of colorectal carcinogenesis and their clinical pictures. *Ann N Y Acad Sci* **910**, pp. 10-8; discussion 18-20.

Tomlinson, I. and Bodmer, W. (1999) Selection, the mutation rate and cancer: ensuring that the tail does not wag the dog. *Nat Med* **5**, pp. 11-2.

Tomlinson, I. P., Novelli, M. R. and Bodmer, W. F. (1996) The mutation rate and cancer. *Proc Natl Acad Sci U S A* **93**, pp. 14800-3.

Tsuda, H. and Hirohashi, S. (1995) Identification of multiple breast cancers of multicentric origin by histological observations and distribution of allele loss on chromosome 16q. *Cancer Res* **55**, pp. 3395-8.

Turhan, A. G., Lemoine, F. M., Debert, C., Bonnet, M. L., Baillou, C., Picard, F., Macintyre, E. A. and Varet, B. (1995) Highly purified primitive hematopoietic stem cells are PML-RARA negative and generate nonclonal progenitors in acute promyelocytic leukemia. *Blood* **85**, pp. 2154-61.

Tyner, S. D., Venkatachalam, S., Choi, J., Jones, S., Ghebranious, N., Igelmann, H., Lu, X., Soron, G., Cooper, B., Brayton, C., Hee Park, S., Thompson, T., Karsenty, G., Bradley, A. and Donehower, L. A. (2002) p53 mutant mice that display early ageing-associated phenotypes. *Nature* **415**, pp. 45-53.

Uhr, J. W. and Marches, R. (2001) Dormancy in a model of murine B cell lymphoma. *Semin Cancer Biol* **11**, pp. 277-83.

van de Wetering, M., Sancho, E., Verweij, C., de Lau, W., Oving, I., Hurlstone, A., van der Horn, K., Batlle, E., Coudreuse, D., Haramis, A. P., Tjon-Pon-Fong, M., Moerer, P., van den Born, M., Soete, G., Pals, S., Eilers, M., Medema, R. and Clevers, H. (2002) The beta-catenin/TCF-4 complex imposes a crypt progenitor phenotype on colorectal cancer cells. *Cell* **111**, pp. 241-50.

van Dekken, H., Vissers, C. J., Tilanus, H. W., Tanke, H. J. and Rosenberg, C. (1999) Clonal analysis of a case of multifocal oesophageal (Barrett's) adenocarcinoma by comparative genomic hybridization. *J Pathol* **188**, pp. 263-6.

Vogelstein, B., Lane, D. and Levine, A. J. (2000a) Surfing the p53 network. *Nature* **408**, pp. 307-10.

Vogelstein, B., Lane, D. and Levine, A. J. (2000b) Surfing the p53 network. *Nature* **408**, pp. 307-10.

Waldman, F. M., DeVries, S., Chew, K. L., Moore, D. H., 2nd, Kerlikowske, K. and Ljung, B. M. (2000) Chromosomal alterations in ductal carcinomas in situ and their in situ recurrences. *J Natl Cancer Inst* **92**, pp. 313-20.

Wang, X., Jin, D. Y., Wong, Y. C., Cheung, A. L., Chun, A. C., Lo, A. K., Liu, Y. and Tsao, S. W. (2000) Correlation of defective mitotic checkpoint with aberrantly reduced expression of MAD2 protein in nasopharyngeal carcinoma cells. *Carcinogenesis* **21**, pp. 2293-7.

Wassmann, K. and Benezra, R. (2001) Mitotic checkpoints: from yeast to cancer. *Curr Opin Genet Dev* **11**, pp. 83-90.

Westervelt, P., Lane, A. A., Pollock, J. L., Oldfather, K., Holt, M. S., Zimonjic, D. B., Popescu, N. C., DiPersio, J. F. and Ley, T. J. (2003) High-penetrance mouse model of acute promyelocytic leukemia with very low levels of PML-RARalpha expression. *Blood* **102**, pp. 1857-65.

Westervelt, P. and Ley, T. J. (1999) Seed versus soil: the importance of the target

cell for transgenic models of human leukemias. *Blood* **93**, pp. 2143-8.
Wijnhoven, S. W., Kool, H. J., van Teijlingen, C. M., van Zeeland, A. A. and Vrieling, H. (2001) Loss of heterozygosity in somatic cells of the mouse. An important step in cancer initiation? *Mutat Res* **473**, pp. 23-36.
Wilkens, L., Bredt, M., Flemming, P., Klempnauer, J. and Heinrich Kreipe, H. (2000) Differentiation of multicentric origin from intra-organ metastatic spread of hepatocellular carcinomas by comparative genomic hybridization. *J Pathol* **192**, pp. 43-51.
Winton, J. D. (2001) Stem cells in the epithelium of the small intestine and colon. In Stem Cell Biology (ed. D. R. Marshak, R. L. Gardner and G. D.). Cold Spring Harbor: Spring Harbor Laboratory Press.
Wodarz, D. (2001) Viruses as antitumor weapons: defining conditions for tumor remission. *Cancer Res* **61**, pp. 3501-7.
Wodarz, D. (2003) Gene Therapy for Killing p53-Negative Cancer Cells: Use of Replicating Versus Nonreplicating Agents. *Hum Gene Ther* **14**, pp. 153-9.
Wodarz, D. (2004) Checkpoint genes, ageing, and the development of cancer. *Oncogene* **23**, pp. 7799-809.
Wodarz, D., Iwasa, Y. and Komarova, N. L. (2004) On the emergence of multifocal cancers. *Journal of Carcinogenesis* **3**, pp. 13.
Wodarz, D. and Jansen, V. A. A. (2001) The role of T cell help for anti-viral CTL responses. *Journal of Theoretical Biology* **211**, pp. 419-432.
Wodarz, D. and Krakauer, D. C. (2001) Genetic instability and the evolution of angiogenic tumor cell lines (review). *Oncol Rep* **8**, pp. 1195-201.
Wong, W. M., Mandir, N., Goodlad, R. A., Wong, B. C., Garcia, S. B., Lam, S. K. and Wright, N. A. (2002) Histogenesis of human colorectal adenomas and hyperplastic polyps: the role of cell proliferation and crypt fission. *Gut* **50**, pp. 212-7.
Wright, N. A. and Poulsom, R. (2002) Top down or bottom up? Competing management structures in the morphogenesis of colorectal neoplasms. *Gut* **51**, pp. 306-8.
Yamamoto, H., Min, Y., Itoh, F., Imsumran, A., Horiuchi, S., Yoshida, M., Iku, S., Fukushima, H. and Imai, K. (2002) Differential involvement of the hypermethylator phenotype in hereditary and sporadic colorectal cancers with high-frequency microsatellite instability. *Genes Chromosomes Cancer* **33**, pp. 322-5.
Yatabe, Y., Tavare, S. and Shibata, D. (2001) Investigating stem cells in human colon by using methylation patterns. *Proc Natl Acad Sci U S A* **98**, pp. 10839-44.
Yee, K. W. and Keating, A. (2003) Advances in targeted therapy for chronic myeloid leukemia. *Expert Rev Anticancer Ther* **3**, pp. 295-310.
Yoon, D. S., Wersto, R. P., Zhou, W., Chrest, F. J., Garrett, E. S., Kwon, T. K. and Gabrielson, E. (2002) Variable levels of chromosomal instability and mitotic spindle checkpoint defects in breast cancer. *Am J Pathol* **161**, pp. 391-7.
You, L., Yang, C. T. and Jablons, D. M. (2000) ONYX-015 works synergistically with chemotherapy in lung cancer cell lines and primary cultures freshly

made from lung cancer patients. *Cancer Res* **60**, pp. 1009-13.

Index